REALISM:

The INTEGRATION of SCIENCE and PHILOSOPHY.

Copyright © 2022

Uwe C. Koepke

"All rights reserved. No part of this book may be reproduced or transmitted in any form or by any means, electronic or mechanical, including photocopying, recording, or by any information storage and retrieval system without the written permission of the author, except where permitted by law."

ISBN   979-8-9861169-2-1

Cover photo source credit: egil sjøholt

Dedication

In grateful memory of:

My mother Gertrud and grandmother Franziska who brought me to America for greater opportunity in the post WW II years.

My mother-in-law Charlotte Mauks and father-in-law Charles Mauks MD who made my long education possible.

"The consensus seems to be that philosophy is currently at a low ebb. ... What is true is that academic philosophy has become rather stale. It is obsessed with its own past, suspicious of radical new insights, inward looking, largely removed from worldly concerns, and therefore of hardly any help in tackling most of the issues faced by ordinary people. Hence the word *crisis* in the title of this book."

Mario Bunge
"Philosophy in Crisis: The Need for Reconstruction", 2001

TABLE OF CONTENTS

Pages

ix   Frequently Used Abbreviations

x   Preface- Scientific Realism: Philosophy in the Light of Science

xii Acknowledgements

1   Introduction: Science-based Philosophy- A Worldview

5   Prologue: Some Needs to Achieve Progress
P.1 Is progress in philosophy possible?
P.2 What is wisdom?
P.3 The place of the scientific method (science) in philosophy
P.4 What of "philo" in philosophy?
P.5 Concepts, words and language – the need for clarity
P.6 Wisdom and the 'purpose of life
P.7 A definition and the usefulness of philosophy

15 Part I – Ontology: The Basic Building Blocks of Reality

15 Chapter I – Matter/Energy, the P-Laws and Emergence
1.1 The "Laws of Physics" In A System of Secular Hypognosticism
1.2 The Early Greeks
1.3 Standard laws of physics plus emergence: The P-Laws
1.4 Why Include Emergence and Why It Is So Important
1.5 What Are the Emergent Properties of Water?

27 Chapter II – Emergence of Life and Mind
2.1 How Emergence and Ontology Allows Life & Mind
2.2 Brain-Mind
2.3 Structure and Organization
2.4 Life on the Sun – A Telling Misconception
2.5 Beyond Water and Inorganic Chemistry —Organic Chemistry.
2.6 Why Not Silicon Instead of Carbon?
2.7 A Closed Environment — Bubbles
2.8 Energy
2.9 The Conservation of Structure: From Nano- To Micro -Structures To Convergent Evolution
2.10 Possibility, Probability, Property, and Randomness
2.11 A Short Excursion into Mathematics
2.12 Probabilities in "Bubbles" Revisited
2.13 Knowledge as a Tool
2.14 Reality, And Its Models – A Preview of Epistemology

43 Chapter III – Possible Pathways to Life
3.1 Connecting Pre-Biologic Systems and Life
3.2 Repair and Maintenance – The Importance of Further Organization
3.3 Defining "Life"
3.4 The Emergent Properties of Life
3.5 Cellular Differentiation
3.6 Growth and Early Reproduction
3.7 From RNA/DNA To Organelles
3.8 Organ Formation and the Nervous System
3.9 Sexual Reproduction in Eukaryocytes
3.10 Advantages of Multiple Organs
3.11 The Neuron — A Special Kind of Cell
3.12 The Transformation of Information by Analog and Digital Means

55   Chapter IV - Defining Brain-Mind

4.1 The Ontology of the Brain-Mind.
4.2 Knowledge- the Emergent Possibility of Mental Models
4.3 Scientific Studies Concerning the Nervous System,
– An Overview
4.4 Why "Typical" and not "Normal"
4.5 Vocabulary and Semantics in Philosophy: A Form of Information.
4.6 The Meaning of Meaning
4.7 Causation: A Physical Concept
4.8 Unknowability – the Root of a Hypognostic Outlook
4.9 Neuronal Sub-Systems
4.10 Mind — A Process
4.11 Nervous Systems (NSs) That Function as a Brain- Mind
4.12 Memory -Laying Sown and Retrieval
4.13 Divisions of the Brain and its Effects on Philosophy

73   Part II – Epistemology: The Study of Knowledge

Knowledge: A Temporary Conscious Model

75   Chapter V Knowledge: A Mental Model
5.1 The Connection with Ontology.
5.2 What Does it Mean to" Know"
5.3 Knowing and Surviving
5.4 The Lay Uses of the Word "Know"
5.5 The Relationship of Epistemology to Ontology
5.6 Structure of the Nervous System, Continued

5.7 Building A Mental Model. Abstraction – Definition and Example
5.8 Vision as An Example of Abstraction
5.9 Imagination and Confabulation
5.10 The fMRI – A Tool for Studying Brain Function "In Vivo"
5.11 Levels of Brain-Mind Complexity
5.12 What Can We "Know?"
5.13 Belief, Faith, Certainty, Knowledge and Fact
5.14 Fantasy – the Basis of All Models
5.15 Consciousness – An Analogue Process
5.16 What Can We Know?

93 Chapter VI - Truth - and "The Thing in Itself"
6.1 The Kinds of Truths
6.2 Finding Conceptual Truth
6.3 Received Truth
6.4 A Return to a Defendable Quest for Understanding
6.5 The Limits of Knowledge: The Case for Hypognosticism
6.6 Truth and the Lay Uses of the Verb "To Know"
6.7 Can Knowledge Be Received?
6.8 What to Do When Authorities Disagree
6.9 Culture and the Tribe as Authority

109 Chapter VII- Concepts
7.1 The Development of a Concept
7.2 Ontology of a Concept - the Basis of Epistemology
7.3 The Energies of Mysticism
7.4 What Do Plants, Animals, Infants and Children Know?
7.5 What Is Consciousness?
7.6 What Is Wisdom?
7.7 NLTs, RLTs and MLTs – Categories of Being

121 **TABLE I** - Characteristics of the interactions of existents

125   Chapter VIII - Wisdom and its Needs

8.1 Communication and Wisdom
8.2 Words and Their Concepts
8.3 What Is Meaning in Communication?
8.4 How to Deal with Changing Meanings of Words
8.5 The Purpose of Wisdom
8.6 Creating Purpose in A Hypognostic State

135   Part III Evaluations
135   Chapter IX – Beyond Reflexes and Instincts
9.1 The Emergence and Influence of Thought
9.2 Wisdom –A Product of Evaluation
9.3 Why "Evaluations" Rather Than "Metaphysics" Or "Axiology"
9.4 Ontology of Mind – Continued
9.5 A Newer Level of Emergence — Self-Awareness
9.6 Is Life Sufficient for there to be Evaluative Functions?
9.7 The Emergence of the Evaluative Ability of Physiological Needs and their Fulfillment
9.8 Axiology and the Determination of Value
9.9 Defining Intentionality and Subjective Meaning: Some Models of Neurologic Activity
9.10 Thoughts on Evaluations Based on Psychological Needs
9.11 What are the Kinds of Evaluations Performed by a Nervous System?
154   **Table II –** The Evaluative systems
9.12 What Underlies the Evaluations of Human Interactions?
9.13 The Ever-Changing Individual Brain-Mind
9.14 A Way Forward

161 Chapter X – The Concept of Value
   10.1 *The Cognitive Basis of Value*
   10.2 *Preferences – the I-E Values*
   10.3 *Value of life: Process Versus a Specific Instantiation*
   10.4 *Identity*
   10.5 *The "Equal" Value of a Living Thing*
   10.6 *The Trolley Problem – a Popular Thought Experiment*
   10.7 *"Pure" I-E Goals?*
   10.8 *Emergence of the Concept "Future"*
   10.9 *Beyond the Preconscious Adaptive Mechanisms (PAMS)*
   10.10 *Conscious Evaluations Are Serial*
   10.11 *The Parallel Evaluations*

179 Chapter XI – The Function of Purpose
   11.1 *The Expansion of Imagination and Memory*
   11.2 *Universal and Personal Goals*
   11.3 *A Short Discussion of the Limits of Near-Common Universal Purposes*
   11.4 *What Are Achievable Personal Purposes?*
   11.5 *The Ethical Purposes of a Life*
   11.6 *Aristotle and Goals*
   11.7 *Happiness and Welfare*
   11.8 *Values from Mental Harmony*
   11.9 *A Life Worth Living*
   11.10 *The Perpetuation of Life*
   11.11 *How to Evaluate Harm and Destruction*

191 Chapter XII – The Two Evaluative Systems and Aesthetics
   12.1 *The Instinctual-Emotive (IE) and the Conscious-Logical (CL) Systems: A Contrast*
   12.2 *The Logic of Aesthetics: What Does Beauty Have To Do With Wisdom?*

12.3 Aesthetics — the Relative Importance of an Emotional and Instinctive Evaluation
12.4 Evaluations Beyond the Aesthetic
12.5 Instinct and Emotion Are Often Right
12.6 The Senses
12.7 Olfaction, Taste and the Tactile Senses
12.8 Vision
12.9 Hearing.
12.10 The Beauty of Logic and Art

198 Part IV – Toward Wisdom:
The Application of Knowledge, Values and Purpose
Ethics: "Do Right Thing"

201 Chapter XIII – The Concept of Ethics
13.1 Ethics –An Overview
13.2 Working Definitions for Ethics: Means, End (Goal) and Purpose
13.3 Different Intended Actions: Good (Benevolence) Versus Evil (Harm)
13.4 *Medical Decisions and Ethics*
13.5 The Effect of Distance and Power
13.6 Expanding on Definitions as They Pertain to Ethics
13.7 The Place of Symbiosis – an Example of a Muddy Definition
13.8 Primacy of the End Over Means in Ethics
13.9 The Central Place of "Should" In Ethics
13.10 Humane Pragmatism
13.11 On the Concept of Free Will: Statistical Randomness or Cause-Effect Relationship
13.12 The Extent of Free Will and the Concept of Evil

225 Chapter XIV - The Source of Human Ethics
14.1 Needs, Values and Satisfiers in Ethics – A Review of the Complexities
14.2 An Ethics Based on Needs and Therefore Values
14.3 The Needs and the Human NS Methods of Satisfying Them
14.4 *Balancing Needs in A Variable Environment*
14.5 Typical Shared Human Purposes Leading io Ethics
14.6 Limits on a Common Secular Humane Ethics
14.7 The Neurologic Conflicts of Means and End
14.8 The Impossibility of an Absolute Ethics
14.9 Ethics and Wisdom

241 Chapter XV - Justice as Applied Ethics
15.1 Introduction
15.2 The Pursuit of Justice
15.3 The Difficult Path to A Universal Humane Ethics and Justice
15.4 Subtypes of Applied Ethics: Medical, Legal, Business , and Other Groups
15.5 Levels of Power, Its Effects and the Subsequent Application of Justice
15.6 Life and Death
15.7 Purposeful Death: Capital Punishment, War, Abortion . Euthanasia, Lethal Neglect
15.8 Is an Ethical Response to Evil Power Wise?

260 Epilogue: A Rational Hope

261 Glossary

293 Bibliography & Resources

FREQUENTLY USED ABBREVIATIONS IN THIS BOOK

C-L = Conscious – Logical; see also I-E

DNA= Deoxyribonucleic acid, the blueprint of a species.

H. sapiens = the human species *Homo sapiens*

I-E = Instinctual-Emotive; see also C-L

LTM =Long-term memory

NS = Nervous System

PAMs = Preconscious Adaptive Mechanisms

PFC = Prefrontal cortex of the brain

P-laws = the axioms and derived laws of physics including emergence

§ = Section; §§ = Sections

STM = Short-term memory

SEP = Stanford Encyclopedia of Philosophy

    (www. https://plato.stanford.edu)

    Types of BEING:

MLT = MODELing Living Thing

NLT = Non-Living Thing

RLT = Reactive Living Thing

PREFACE

Scientific Realism: Philosophy in the Light of Science

This work is the result of addressing some of the problems noted in the book "Philosophy in Crisis" (Bunge, 2001). It is meant to appeal, at an approachable level, to students (both formal and self-taught) of both **science** and philosophy. It deals with a form of 'scientific realism', a worldview which deals with the observed, and in **theory** observable, properties of our **universe** (Bunge, 2006). Science and philosophy are two related methods that **human**s have used to understand our world. Philosophy, as a field however, has not developed a general **consilience**[1] with our scientifically gained understanding of the parameters of reality. This book is not about philosophy of science, the effort by (mostly) non-scientists to define what the concepts concerning science **should** be based on. This book reverses that stance, toward how philosophic concepts should be informed by scientific ones. Another principle, namely the basic **principle** of non-contradiction, is also basic to scientific realism. This principle was already formulated and discussed by Aristotle. (*Aristotle on non-contradiction*), *Stanford Encyclopedia of Philosophy (SEP)* on line: *https://plato.stanford.edu/entries/aristotle-noncontradiction*). Concepts that do not have any foundation in common or confirmable human experience (direct or indirect), or that contradict those that do, are considered, provisionally, as personal exercises of the **imagination** and are central to **Idealism** philosophies. Unsupported or unsupportable concepts that are clearly impossible constructs when held up to the light of science will be held in abeyance or discarded. The book

---

[1] *The first use of a word that is the symbol of a central concept will be in bold print suggesting a use of the Glossary. There the specific meaning for that word, as used in this book, will be found. This list is based on dictionary and encyclopedia entries, the work of Bunge (2003) and Mead (1946) as well as wide reading on the topics.*

as a whole will show how science provides a framework for a modern **ontology** and epistemology. These topics in turn provide the grounding for a consilient discussion of **values**, **ethics** and justice, the **goal** of **wisdom**. The Glossary will be seen to be of great importance because its creation forced the author to define the concepts that have traditionally been used loosely or in **polysemous** ways. This lack of precision has been, and still is, a major factor in the dearth of philosophical progress especially as compared to the sought-after clarity in the scientific fields of endeavor.

Because many readers will have some experience in one area but less in the other, the *Introduction* summarizes the layout of the book so that they receive an overview of the thrust of the arguments and their relationships. This will make the book, with its broad content, much less daunting.

> *"The consensus seems to be that philosophy is currently at a low ebb. ... What is true is that academic philosophy has become rather stale. It is obsessed with its own past, suspicious of radical new insights, inward looking, largely removed from worldly concerns, and therefore of hardly any help in tackling most of the issues faced by ordinary people. Hence the word crisis in the title of this book."*
>
> Mario Bunge
> *"Philosophy in Crisis: The Need for Reconstruction"*, 2001

Acknowledgements

Tania Robie, who showed me the importance of editing and encouraged my early attempts in writing.

J. Gordon Roy, my late good friend, who spent countless hours discussing and helping me clarify my ideas. Covid and his illness ended that collaboration.

Frank J. Dye, PhD for clarifying concepts and vocabulary in cell and developmental biology.

Elizabeth Schneewind, who did the first copy editing of the manuscript.

David Kirkland, who did a critical review.

Ken Brooks, who took my cover ideas and made them better.

And, of course my extended family which encouraged me the last 15 years of this effort. Especially, my wife, Anne Mauks, MD, who proofread the whole book many times as I revised it. Any possible remaining errors are my responsibility.

Introduction

A Science-based Philosophy- A Worldview

The main body of the book begins with a section concerning the general **need**s that should be addressed to foster progress. This is followed by four major parts that cover philosophy under commonly recognized broad subdivisions. They are **Ontology, Epistemology, Evaluations**, and **Ethics** and **Justice**. The first two, in our modern age, are now highly influenced and defined by science. But the latter two areas are not yet influenced enough by the former. Science is a method. Data gained thereby are the building blocks of hypotheses and theories in physics, chemistry, biology, psychology, sociology, and anthropology among others. Together these fields of study have led to a cumulative increase in the understanding of our universe. The processes of the **imagination** do not define the endpoints of understanding, but rather are a means leading toward further investigation.

Although the word 'theory' is used in philosophy as in "A Theory of Justice" (Rawls,1971), the discussion, by most philosophers, of the topics from ontology to justice are not based on the theories from physics through neuroscience. Rather they are based mainly on **introspection**, untethered imagination and **transcendent** ideas. This book shows how philosophy can be correlated with science without aspiring to **absolutes**. This can be done by realizing that our ideas, at their best, are only models of our mind, and cannot mirror reality completely. It also shows that '**truth**' is a fraught term, its **meaning** being highly **subjective**. With this in mind a new word seemed useful, and I introduce the new term, "**hypognosticism**". It acknowledges the unavoidable limits of all knowledge. It accepts that the process of science can only approach a complete understanding of the universe, albeit on the basis of confirmable data. In this, science offers more **certainty** than **agnosticism**. The latter is however included in the concept of hypognosticism, whereas **atheism** and descriptive **theology** are not. It is used as a reminder of our limits in creating **mental model**s and should foster a shield against unwarranted certainty.

The fifteen chapters of the main four Parts are further subdivided into numbered subtopics for ease of cross referencing and to signal slight changes in focus. For example, '*1.3 The **standard laws of physics** plus emergence: The **P-Laws**'* is the third section of chapter one.

Part I concerns **Ontology.** It is presented in a form that is based on the currently accepted laws of physics (herein referred to as the **P-laws).** But it critically adds the concept of **emergence** (Vintiadis**, 2021)**, namely that **structure**, size and complexity add non-metric possibilities to the underlying properties of sub-atomic **matter/energy**. These added properties, including those of **life** itself, are due to unique relationships, especially those due to changes in the level of complexity. In § 4.6 the need for clear **definition**s is also covered. By comparing the relative clarity of scientific terms versus the incompatibility of philosophic terms and their conceptual referents I propose that the field of philosophy seek similar clarity – and how this may be accomplished.

Part II covers **Epistemology,** as based on modern neuroscience research, from anatomy to psychology, and its findings. The arguments show how knowledge is nothing more or less than a mental model based on reality. The **impossibility** of mental models of reality to be identical to reality, the **phenomenon** versus the **noumenon** of Kant, is stressed. How this affects discussions of the evolution of life and mental abilities is presented. Chapter VI discusses the meaning of the term 'truth', a highly fraught term as currently used. There are many subtypes of truth. The concepts concerning the actualities studied by ontology and epistemology are less limited though hypognostic models of reality than those created by the evaluative and ethical processes.

Wisdom, an application of perceived truth, is one of the aims of philosophy, as discussed in Chapter VIII. It concerns the **purpose** and propagation of wisdom as communicated to others, as well as a source of well-considered actions. These chapters form the basis of the of fraught concepts of what can be the sources of **value** (intrinsic and extrinsic), ethics and justice.

The study of **Evaluation**, including the general term "value" is the topic of Part III. Values will be seen to be properties of actions or objects based on their ability to alleviate or worsen needs. But there are two majors, competing, systems of the **brain-mind** that ascertain value. They are the **Conscious – Logical (C-L)** and the **Instinctual-Emotional (I-E)** systems. The latter is much older in evolutionary terms and gives quicker, but less nuanced results than the former. Chapter IX shows how the former process rests on the emergence of a brain-mind of sufficient complexity for the logical aspects of the task. A scientifically consilient system of ethics and justice will be seen to depend, logically, more on the C-L system, although science can give insights as to the source and pros and cons of the I-E evaluations.

In Part IV **Ethics** and **Justice** are discussed in terms of the previous chapters, rather than as a freestanding Platonic idealism exercise. The great conundrums in this area are the inherently different needs of the **self,** versus other individuals, and the even greater problem of how to order the needs of the self with those of one's own group (small to large) or those of others, and finally one's relationship to the greater world at large. Here we deal with the most difficult concepts of philosophy: Good and Evil, Free Will, the controversial issue of ethical values, and the impossibilities of an absolute ethics being major topics. The chapter on Justice highlights the need to use modifying adjectives to create phrases that differentiate each subspecies of justice from the more general family name "justice". Sections on life and death, and the ethical responses to abuses of power round out the last chapter.

The Epilogue offers the **hope** that philosophy can help reach greater agreement, on the basis of scientific models and the logic of ethics and suggest improved reactions to the stresses of this world. In the Anthropocene era the results of the human impacts, on the each other and the world, are now clearly evident. The best minds, educated in scientific and philosophical concepts, will be needed to come up with a positive response to this crisis. This might lead to a more promising future than is currently feared. This book is an effort in that direction.

Prologue:
Some Needs to Achieve Progress

P.1 Is progress in philosophy possible?
Philosophy has been in a crisis for many years (Bunge, 2001). There have been papers at the margins that indicate progress in meeting the ethical needs of humanity, but most writings have been discussions of the earlier works written in an **environment** of very limited scientific understanding. The rapid increase in improved models of reality as investigated by science has not been incorporated in the thinking of most professional philosophers. Academics from **mathematics**, neuroscience, biology, anthropology and so on have made important contributions, but this has not filtered down into the **standard** academic curriculum of the majority of colleges and universities where linguistics, logic and historical analysis predominates.[2] Of note is that scientific fields are not mentioned. If philosophy is to survive as an academic field, and not follow the path of alchemy and astrology the undertaking needs to progress.

---

[2] "Common Coursework Philosophy Majors Can Expect

Philosophy majors should expect their degree requirements to cover some fundamental topics while also leaving space to explore. Majors can take classes that survey parts of philosophy's history, provide an introduction to logic and explore the philosophy of language. They will also likely become familiar with some new vocabulary, as parts of the field like metaphysics and value theory drive many class options and requirements. Broadly, metaphysics concerns the fundamental nature of the world, while value theory includes moral, social and political philosophy. By taking classes that host discussions for difficult questions and provide many possible answers, philosophy majors will become skilled at forming arguments that are critical and reasoned."
Pincus, Melanie, *"What You Need to Know About Becoming a Philosophy Major"*, https://www.usnews.com/education/best-colleges/philosophy-major-overview May 5, 2020

I define progress in philosophy as an improvement in the mental models and concepts that mirror reality. Such better understanding, will lead, in turn, to better predictions of how and which means have the highest **probability** of achieving an improved commonly valued goal. This understanding, and the ensuing actions will contribute to a definition of "wisdom" as based on "Sophia", the second root of the word "philosophy". What science can contribute is then laid out in Section P.3. The emotional relationship, the "philo", of the philosophers toward their subject, will also be discussed below. Speaking broadly, philosophy can be thought of as "the **desire** and urge to acquire and act with wisdom". The bases and possible improvements that can lead to such progress is the subject of the rest of this book.

P.2 What is wisdom?

To understand what philosophy might refer to we will need to inspect the constituent parts of the word. To that end let us look at what working philosophers and lexicographers have said.

It is indeed surprising that R. Audi, the editor of the "The Cambridge Dictionary of Philosophy, Second Edition" (1999) (CDofP), felt that he could not include an entry under the term "philosophy". The reason for this decision was explained in the introduction:

"Some readers might be surprised to find that there is no entry simply on philosophy itself. This is partly because no short definition is adequate. It will not do to define 'philosophy' in the etymological way many have, as 'the love of wisdom'".

My immediate reaction was, "Why not a long discussion?"

A more accessible source, the "Webster's New Collegiate Dictionary" (1956) gives us five uses of the word.

"1. Literally, the love of wisdom; in actual usage, the science which investigates the facts and principles of reality and of human nature and conduct; specif., and now usually, the science which

comprises logic, ethics, aesthetics, metaphysics, and the theory of knowledge. 2. A body of philosophical principles; of specific major branches of learning...". and continues with other derivative uses.

Mario Bunge in his philosophical dictionary (2003, pg. 213) defines it as follows.

*"**PHILOSOPHY** The discipline that studies the most general concepts (such as those of being, becoming, mind, knowledge, and more) and the most general hypotheses (such as those of the autonomous existence and knowability of the external world)".*

He goes on to list major basic branches, some of which overlap with those listed in Webster's Dictionary.

John Passmore, in the "The Encyclopedia of Philosophy" (1967) under the entry "philosophy," gives no definition, but proceeds, as in all the other entries, by a discussion of the topic. He notes that

"'... 'philosophy' ... is translated as 'the love of wisdom.' But "sophia" had a much wider range of application than the modern English 'Wisdom'".

The article then continues by using the subheadings of: the Platonic conception, the knowledge of ultimates, relationship between philosopher and sage, the idea of a philosophical method, philosophy of value, the philosopher as advisor, philosophy and the **special** sciences, philosophy as the science of man, philosophy as the science of sciences, philosophy as speculative cosmology, philosophy as a theory of **language**, philosophy as the theory of critical discussion, fields of philosophy, "description, prescription, and rational reconstruction", and ends with "the variety of philosophical tasks".

These are subdivisions of various topics that may be addressed by the field, but does not clarify what philosophy offers that is not offered by the physical and social sciences or the humanities. It also does not address whether sophia can arise from **fantasy**.

Can common threads be found in these various definitions and discussions? Or is the task too daunting, as the editor of the CDofP states. I believe it is not only possible, but as for all central terms in a discussion, it is necessary that at least a minimal definition of the word, and its associated concept is given.

Certainly, wisdom is a key concern of philosophy. With this in mind I propose this definition. Wisdom is:

"A product of a <u>mind</u>, based on a highly informed <u>understanding</u> of the constituents and relationships of our world, which in turn leads to refined <u>predictions</u> of which possible actions lead to the highest probability of achieving a specific <u>goal</u> and its <u>means</u>"

This definition of wisdom covers all the basic topics with which philosophy deals. It is complicated to the degree that it includes all the necessary components of the concept. Throughout this book the noun "mind" refers to a process whose **material** bases are a physical substrate <u>and</u> the **energy** that flows through it. To emphasize the living example of this combination the compound word <u>brain-mind</u> is used extensively. Computer processes such as "Artificial **Intelligence**" are not, as yet, comparable with mind as discussed in § 2.2.

Wisdom implies the attempt to bring the best **information** to bear on the problems of life. To this end, scientific concepts concerning the constituents and relationships of our world are covered in Part I, Ontology. What is meant by understanding and knowledge of reality is the purview of Part II, Epistemology. The best predictions, based on the above, as to which actions would have the highest probability of achieving a goal will be seen to involve both ontology and epistemology. But information is always incomplete. This leads to the conclusion that understanding will usually be incomplete. This hypognostic state means that evaluations (Part III to IV) should rarely be presented in absolute terms. A wise **person** also realizes that evaluations involve concepts such as conscious logical determinations as well as the more evolutionarily primitive emotions and the combination of both. Unfortunately, these two important functions, products of the brain-mind, may, or may not, be harmoniously integrated. Only then can the means and goal that are to be (should be?) pursued be wisely formulated. Finally, in Part IV, the study as ethics and justice are treated as the special fields that evaluate the wisdom of actions (means) toward an overriding goal.

P.3 The place of the scientific method (science) in philosophy

The purpose of this book is to discuss philosophical inquiry in the light of the methods used, and the results gathered, by the scientific method. As is common in scientific fields, central words and concepts are discussed in turn in the effort is made to make all the definitions fit into a consistent and consilient framework. Furthermore, the various conclusions arrived at will also be held to a consilient standard, in concordance with the best available evidence. Absolute statements will be rare.

Science can be seen as the child of philosophy and has been asked by some to replace its parent. Such parentage refers to the attempts of early philosophers to answer questions concerning the basis of reality and knowledge. The answers were based on **logic**, but very little data other than that available directly to the unaided senses. But a full replacement cannot be undertaken because the methods of science cannot deal with the concept of '*should*'. The concepts of science can however underpin progress of philosophy. Science primarily seeks details of the what and the how of reality. Philosophy, at its best, uses these details and then coordinates, it is to be hoped, the whole breadth of human understanding. The information garnered by science will never be sufficient, standing alone, to provide interpretations of topics such as value and ethics. These have strong emotional components, and emotions are not the result of mental mechanisms that drive scientific inquiry. How to make a sausage, what its components are, where it might be found, and what are its nutritional consequences, are questions that can all be answered in a scientific way. Whether sausages *should* be made at all and for what **potential** goal they *should* be used are not questions answerable by science. This is because *should* invokes a different order to our purposes than the hierarchy of their probability or possibility of actualization. The discussions in the chapters covering ethics discusses the underpinnings of this conclusion.

However, no matter what field of inquiry, ignoring the data and information garnered by **objective** means often leads to unfounded and potentially unbounded conclusions such as is seen in fantasy. Science insists that there are bounds to possibilities. Concepts which are arrived at by willfully ignoring well-established bounds lead only to accidental, rather than demonstrable, wisdom. **Ignorance** is the lack of exposure to knowledge and is expected and forgivable. Willful ignorance, or "**wignorance**" (my portmanteau word), is anathema to science, and should be decried in philosophy, as elsewhere. Philosophy can be more than shooting imagined arrows at imagined creatures in the dark, hoping for a meal, even if the meal is only meant to be "food for (philosophical) thought".

P.4 What of "philo" in philosophy?

The interpretation of the first part of the word, "philo", is even more contentious than describing what wisdom might be. It involves a form of "love ", in this case a relationship between the self and pertinent knowledge that leads to wisdom. Unfortunately, any such interpretation requires a definition dependent on the findings of the very field that it is trying to define. However, at the very least, we can say that the first part of the word expresses an affinity, rather than an aversion, between the brain-mind of an individual and wisdom. The discussion of **possible** relationships between the emotional self and the act of making mental models of the world will be seen to coordinate such concepts as free will, purpose and meaning with our scientific understanding.

P.5 *Concepts, words and language – the need for clarity*

It behooves the philosopher to emulate the scientist in aiming for clarity of concepts and their associated words. To this end this book contains a Glossary. It is a list of definitions of central or new (to the reader) concepts. It also gives references to the sections in the book where there is a discussion of the use of the term. An excellent example of a glossary in a science book is that found in a book on human development (Dye, 2000). It contains xxxx entries for a subfield of mammalian biology, all for the purpose of clear **communication** and consilient learning. To emphasize this need there is a discussion of language; language being broadly defined as any method of purposeful communication. This discussion is found in § 4.5 under the heading "*Vocabulary and Semantics in Philosophy: a form of information*". The meaning and impact of the important concept of emergence, treated in chapters I, II, IX and X, is but one example of the needed quest for clarity. Understanding this concept shows that the simple reduction of relationships to their additive effects, the fallacy of reductionism, is a" red herring". This fallacy pervades much of the arguments in the discussions pertaining to evolution, cause and effect, as well as the complexities of the brain-mind.

P.6 Wisdom and the 'purpose of life'

In part IV, the discussions concerning purposes of life (or anything else) will be based on earlier discussions of value, the good and the bad, of helpfulness and evil.

Wisdom implies that appropriate purposeful actions and reactions, based on conscious thought which itself is ultimately based on internal and external **stimuli** (raw information), are attainable. Internal stimuli are those that consists of the interactions of many active brain circuits. These would include **emotion**s, imagination, memory and other circuits that are in the **unconscious** realm. The external stimuli, by definition, are limited to those stimuli whose initial energy source is outside the physical structure of each individual being. The purpose of one's life in this context is to imagine and work toward a future for the self and the world. Philosophy, based on science, can be a major means toward choosing a wise goal. The Epilogue suggests that actions, implemented on a broad scale and with the use of wisdom, may lead to the hope of a desirable future.

P.7 A definition and the usefulness of philosophy

Altogether these discussions point toward a definition of philosophy that can be stated as follows: *Philosophy is a desire to acquire knowledge, and to use this knowledge so that actions will be wise and in conformity with the goal and values of the individual and one's society, limited only by the possibilities inherent in the basic ontological building blocks and those possibilities emerging from their interactions and combinations.*

The influence of philosophy on the future of humanity is therefore increased or diminished by the unity or disunity of its practitioners. In contrast, the scientific endeavor has been more successful in this regard because its practitioners work toward agreement of its purposes, methods and vocabulary. Science can therefore be used as a guide to the philosophical endeavor, while at the same time supplying pertinent information about this world, including our brain, and its function, mind. This scientifically garnered information is critical in a discussion of ontology and epistemology, and therefore important, indirectly but substantially, to discussions of value, ethics, free will, and so forth. Wignorance will not do. The extent of the influence of philosophy will also be increased if a way can be found to realize the hope that "the pen is mightier than the sword". Progress in the human condition can only occur in the long run if the **power** of wisdom can better control the power of pure emotion that drives so much of human actions.

It may have been noted that there is no section on metaphysics, a term even more fraught in use than philosophy. Metaphysics is often the name given to a third section in a general book on philosophy. In this book its main concerns are discussed in Part I, Ontology. Historically, it was first used by a *"first century, C.E., editor who assembled the treatise we know as Aristotle's Metaphysics out of various smaller selections of Aristotle's works"*. (SEP, 2021). In essence it only referred to those subjects that were discussed after his 'Physics', or natural world, section. Later it tended to include the various subjects that did not clearly fit under the heading of epistemology, logic, or axiology. Metaphysics classically also includes the discussion of concepts derived from axioms that are based on proposed transcendental realms. This subset involves, among other things, the **world view** of **religion**s. These axioms, incorporating a non-physical world view, are not consilient with the scientifically based ontology of this book, and are therefore not investigated further. If transcendent entities are invoked, then anything may be possible, valuable and ethical. It leads to intellectual and emotional chaos.

In summary, philosophy needs to be based on a good understanding of the basics of a realistic ontology, epistemology and how they influence the creation of values, ethics and justice.

Part I – Ontology

Chapter I
Matter/Energy, the P-Laws and Emergence

1.1 The "Laws of Physics" in a System of Secular Hypognosticism
There are many systems of ontology and they have many names. Bunge's system is "Scientific Realism", (Mahner (2001). This is the concept that the whole universe can in principle be described as consisting only of Matter/Energy and its inherent properties. In practice any description of any system is incomplete due to lack of completeness of information. This may be called "Secular Agnosticism. In conjunction with theological Agnosticism, it forms the concept of HYPOGNOSTICISM, a neologism. It supports a skeptical worldview. Another reason for its use is the inclusion of emergence as one of the sources of natural properties. And, from this point forward the laws of physics plus emergence will be referred to as the P-laws (See § 1.3) to distinguish them from all other rules and laws. Secular Hypognosticism addresses that vast area between omniscience and absolute ignorance.

Hypognosticism therefore refers to the limits of knowledge concerning 'the real' versus proposed transcendent (See Glossary for definition used) entities. It is the admission that omniscience is not available to us as human beings (actually to any entity of this universe) and the assertion that we do have workable, predictive mental models (also called understanding) of some of the entities and their relationships in this universe. The term agnostic will therefore be reserved for the avowed ignorance of anything transcendental. Atheism, though not further discussed, is the avowed omniscient certainty that the transcendental realm, if possible, contains no being that resembles the human vision of a god.

One can find oneself arguing in circles when trying to describe the world we live in. This is the common way to express what is called "begging the question" in philosophical discourse. For that reason, philosophers and scientists have made use of the laws of physics (the P-laws) which are themselves based on the axioms of mathematics that underlie the understanding of the P-laws. Axioms, that is the unprovable givens of this field, will be best known to most readers from high school geometry. The axioms of mathematics and logic are statements that are internally (to the system), usefully and self-evidently true. In Euclidean geometry examples such as "two points describe a straight line" and "a circle is defined by its center-point and its radius" are held to be true. Mathematicians such as Riemann and Lobachevsky showed that this was not true in non-Euclidean space. Einstein's space-time proves to be curved, and his calculations go beyond Euclidian axioms. This is the difference between an easily understandable ideal, which is a Platonic geometry, and the workings of the universe where gravity seems to be everywhere and is a factor in how "straight" lines behave in a gravity perfused universe. This leads to the question of whether to deal with our world in a Platonic, that is ideal way, or in a way that takes a more seemingly messy and complex path. There are two reasons why a pseudo-platonic method is best suited in, for example, both philosophy and engineering. For practical reasons it is impossible to apply all the facts of all the interactions that are pertinent — especially at infinitesimally small and infinitely large scales. Compared to the limits of the human mind, modern computers have helped significantly in managing enormous amounts of information. But current computers also have limits, weaknesses and gaps compared to the human mind (See Part II, Epistemology). Fortunately, in the range of time scales and sizes that are pertinent to our everyday lives, and the processes of human civilizations, we can typically deal with a subset of reality. This subset, in terms of size, encompasses the range of electrons to galaxies, and in terms of time deals with nanoseconds to eons. It is when one approaches the realm of the infinitely small or large that one needs to doubt the reliability and applicability of the P-laws. Those edges of finitude, should make us all — if we are honest with ourselves — hypognostic. This is the realization, that only a part of the universe can be appreciated, thought about, understood and described at any one time. But of this partial realm we can make models that are excellent predictors to how things really are, relate and behave. The scientific endeavor has pushed us ever closer to those edges and has shown us the vast complexities inherent in the realms accessible to us. The great strength of science is that it is an organized gathering of confirmable data subject to recurrent reevaluation. Its P-laws allow the creation of consilient models of reality from this data. Finally, this process increases the likelihood that our lives are led by a derived wisdom deserving of the name. Such wisdom must always be tempered by an acknowledged hypognostic mindset, one that does not fill the unknowable gaps with unsupported and non-consilient fantasies.

1.2 The Early Greeks

The ancient Greeks taught, among other lesser-known formulations, that all existence was based on earth, water, air and fire. Some like Democritus also formalized an "atomic" theory around 465 BC. The term atomic meant uncuttable or non-divisible. It is a form of MONISM. These descriptions were not wrong, but rather incomplete because of a lack of information; a simple, era specific, ignorance. These thinkers did not avoid facts, but rather they had no access to information beyond that provided by their unaided senses. These models could only be applied to reality in the most general terms. Their use in specific cases were often quite misleading and therefore led to error.

The building blocks, consisting of earth, water, air and fire, were bases used in an attempt to understand the universe. It can easily be seen how those early basic materials can correlate with a more modern understanding concerning the basis of reality.

> Earth — stood in for what we now call solid matter
> Water — stood in for what we now call liquid matter
> Air — stood in for what we now call gaseous matter
> Fire — stood in for what we now call energy

It was a description based on the most obvious properties of everything with which we come into contact. It was thought that each thing has one of these forms — or some combination thereof. Much thought went into which combinations described which reality – but in the end all had to be relatively simplistic. All the possible combinations (not permutations- whose number is significantly larger) lead only to the basic four and six possible other forms. The total of ten forms cannot help but be inadequate in describing the hundreds of forms of which we are aware, even with only our unmodified senses.

What we now call atoms are not the indivisible, smallest elements intuited by Democritus and others. But the idea that some very small constituents form the bases, the building blocks, of everything else, is still alive and well in particle physics. Whether the final answer will ever be known cannot be answered at this time. That level of reality may be too close to the infinitesimally small for us to perceive, no matter what microscopic (- or mathematical) tools we may develop. Again, we need to be hypognostic, to acknowledge the limits of our current capabilities of gathering the pertinent data. However, despite these limitations, we should not think that we can only aspire to a mental state of being that implies that it is futile to concern oneself with the future, the present and the past of ourselves, our culture, our earth and the larger universe.

1.3 The Standard Laws of Physics Plus Emergence: The P-Laws

Although we now use the word 'atom' for the basic unit of any element (i.e., oxygen, hydrogen, carbon, etc.), the quest for the ultimate indivisible basis of all reality has, as yet, led only to unmeasured, hypothetical units such as "strings". A string may not be the smallest thing, but it is the smallest constituent that has among other properties mass and energy. And it allows a mathematical explanation of gravity. We also speak of a quantum, the smallest possible discrete unit of any physical property, such as energy or matter.

When physicist speak of a quantum of light, a photon, this includes the understanding that it can only have certain properties. Photons themselves seem to be comprised of smaller entities, but none of them will have the same qualities as a photon (or vice versa). This introduces the concept called "emergent property", which will be discussed in some detail below. Emergence refers to the basic concept that combinations of entities and forces, can have new properties that are not inherent or possible for their constituent parts. For an extensive multidisciplinary discussion of emergence and the Theory of Levels see Emmeche (1997). A concept of emergence is a necessary addition to the standard set of P-laws if we are to understand the immense variety of properties we perceive. It is not a classical physical law in that it cannot be mathematically described, but it is clearly needed to understand the result of phase and property transitions that result from many matter/energy interactions (Carrol 2016). The modern concept goes back to the little-known formulation of Nicolai Hartmann (1882-1950), the "categorium novum". He

Matter/Energy, the P-Laws and Emergence

was very active in the development of the ontology of categories as discussed in the Stanford Encyclopedia of Philosophy (SEP) (https://plato.stanford.edu) (2021).

The unsupplemented laws of physics are of much longer standing. The concepts of space, time, matter, and energy undoubtedly go back to the time when our forebears first began to have a model-making nervous system. The evolution that leads from atoms to simple, unicellular life forms and eventually to humanity's highly complex nervous system is dealt with more fully later in this chapter and then in the discussion of epistemology. These processes cannot be understood without the inclusion of emergence.

Modern physics, especially at the limits of particle physics and astrophysics, continues to fine tune both information and understanding of the gathered data. Philosophers, if they want the field of philosophy to deal with reality, must be aware of these changes in the science-based models concerning the properties of reality. If they choose not to, then they may be accused of wignorance, or dealing in fantasy. It is not essential that the philosopher knows or understands the equations or even experiments of physics and the like. It is not necessary to know all the supporting data. However, one must be well acquainted with the resulting principles of being — and therefore the basis of the P-laws and the logical consequences and limits they imply

The current P-laws are bound by these principles. They are not the same as mathematical axioms, but the latter noun is often used outside that field to indicate a basic assumption. The principles of science are derived rather than given a priory as axioms are in mathematics. Principles point to assumptions that limit the possibilities of a system based on them. Some are:

    1. The laws of physics of this universe (P-laws) are the same throughout this universe.

    2. All P-laws limit the possibilities inherent in space, time and matter-energy. Corollary: Space, time and matter/energy are models by the use of which we interpret those P-laws

    3. Our universe is self-contained and unchanging in total mass-energy.

    4. All possibilities of which kind of quanta can exist are inherent in and limited by the first three principles.

    5. The spatial and temporal relationships of the quanta and all higher constructs thereof underlie the emergence of new possibilities and probabilities of being, which in turn are limited by principle #4.

A few clarifications are necessary if one is to understand these principles correctly. By speaking of "our universe" we allow the possibilities of other universes. These are the parallel universes and multiverses inherent in some of the mathematical descriptions of our universe. How – or whether – these proposed universes relate to ours is highly uncertain. This lack of understanding is only an admission of necessary ignorance, an acknowledgment of hypognosticism. However, it seems these other universes cannot interact with ours because to do so they would have to become part of this universe and its P-laws. If this were not so, then any possible understanding becomes logically impossible. If the P-laws can change or vary inside any one universe, then no description can reflect any eternal truths.

The P-laws can be seen as a result of the consilience of all our knowledge. This word, and concept, has been reintroduced in a book of the same name by E. O. Wilson (1998). In short it is the name for the idea that information from independent sources all leads to the same conclusion. It can be seen as a measure of certitude. It is also the source of questions. When something "doesn't fit", we mean that it is not consistent with the rest of our knowledge and beliefs. The data as a whole then cannot lead to a consilient interpretation. When such new information seemingly cannot be fit into the understanding based on our current models of the P-laws and their logical application to the data, one of three things must be questioned. Either the formulation of the some or all of the P-laws needs to be changed (or even eliminated), or the sources and reliability of all information must be more closely examined, or, most likely, if all the data has equal reliability, our interpretations need to be adjusted. Any change in the P-laws is only viable, if all the equally reliable data leads to new consilient P-laws. Judgement as to reliability of the sources is based, in science, on independent verifiability. If a change in P-laws cannot be made, and the reliability of conflicting data is not demonstrably unequal, then interpretations must be hedged in deference to secular hypognosticism. Then the acquisition of data must continue. Such changes of understanding is called progress, in science. The same can be true of philosophy. The difference is that philosophy does not, in general, deal, with the fine points of subatomic physics, astrophysics, chemistry, or the workings of biological systems, but rather more commonly with concerns relating to the human condition. The human condition is however emergent, as described above, from all the interactions

Matter/Energy, the P-Laws and Emergence

occurring even at pre-human levels of existence. It is that myriad of stacked, overlapping and interacting emergent properties that, if correctly understood, can lead to an appreciation of the connection between the atom and the activity of the human brain-mind. (See § 2.1) The very complexity of the interactions at that human level make it much more difficult to reach consilience in philosophy.

1.4 Why Include Emergence and Why it is so Important

The combination of the gases $(2H_2 + O_2)n$ can be explosive with the addition of a small amount of energy and is not equal in properties to the resulting water, $2(H2O)n$. Note that the kind and number of the elements are the same. This is a very simple example of what occurs with emergent properties as described below. The emergent properties that are possible when one has tens or even hundreds of interacting pieces in relationships can, at best, be guessed at prospectively. Another example of an emergent property is bricks in a pile with mortar on top of the bricks versus properly stacked bricks with mortar in between them. The first is a useless pile, the second is a wall that can stop the movement of some matter and energy and thereby create an enclosure.

The modern concept of emergence is relatively new, but critical, whereas the first four principles are classical in structure and content. Emergence, as used in principle #5, is an idea that has not been commonly considered, and will now be dealt with in depth. I first came across the concept of emergence in the book "Scientific Realism — Selected Essays of Mario Bunge" (ed. M. Mahner (2001). It is also mentioned in less recent books including the very readable book by Hunter Mead (1946). However, it has not been central to most of the typical philosophical writings, ancient or modern. But the concept is critical to understanding much of what has puzzled so many thinkers, including philosophers, over time. Its power resides in the fact that it is not one mechanism or one process, but rather a conceptualization of a certain type of change.

1.5 What are the Emergent Properties of Water?

Because it is so important, I will illustrate this concept as it relates to the connections and relationships that are considered noncontroversial and known by anyone with an interest in how things are. I will start with water. Only then will it be discussed as it impacts philosophy.

On our Earth water is very common, although unevenly distributed. It is essential for life and this makes it a pertinent example. Water, commonly written as H2O, is composed of the simplest atom, hydrogen, and another necessary element for human life, oxygen. Many of us learned in our high school years that water is also called "the universal solvent". The components, hydrogen and oxygen, need not apply for this categorization.

Water owes many of its unique qualities because it is a polar compound. It is a molecule whose electrical charge is asymmetrical in the three dimensions. Molecules are made up of two or more atoms and atoms are made up of protons and neutrons in the nucleus surrounded by electrons in the orbit. Neutrons do not have a charge, and protons have the opposite charge of an electron. The convention that we call the electron negative (-) and the proton positive (+) in charge is of course an arbitrary designation. Anti-matter is made up of neutrons, positrons (+) and antiprotons (-). For reasons inherent in the natural P-laws, there seem to be very few, and then only extremely short-lived, atoms of antimatter. Antimatter and matter cannot exist together and are thought to annihilate each other by converting both kinds of matter into energy. The simplest explanation of why the universe is almost exclusively composed of matter is that for as yet unknown reasons, more matter than antimatter existed initially, and what we are left with is the non-annihilated portions, mostly normal matter. (Antimatter is not just theoretical, and particles as complex as anti-helium have been observed). The importance to philosophy of this, very superficial description of the atoms of this universe, is that it lays the groundwork for the appreciation of the complexity and emergent properties of all things and their relationships. For a more detailed, extensive and readable discussion of subatomic physics and emergence a reading of the recent book by Sean Carroll (2016) would be a good start.

We now return to emergence of polarity and its properties as seen in water. Polarity can be thought of as a nano-magnet, which in fact it is. A magnet made of iron is a macro version based on the interaction of electrons of iron in a special configuration. This emergent property (of an iron magnet) is caused by the atoms being "lined up" by previous exposure to a magnetic field. This non-random configuration causes the piece of iron to have its own magnetic field. The details are not important to philosophy, though of course tremendously important to modern technology. Polarity is an emergent property because it is not found in the constituent parts, namely hydrogen and oxygen atoms in water or in the unorganized arrangement of plain iron. There are many consequences because of this novel condition. It allows, forces and

Matter/Energy, the P-Laws and Emergence

limits how water molecules interact with each other and other molecules, compared to the parent atoms that cannot interact in the same manner in the uncombined state. The reason I write both allows and forces is that the difference between the two is only a matter of probability. The fact that the new interactions are allowed indicates that the emergence of the new potentialities have a probability greater than zero. Furthermore, the new properties, are not arbitrary or a matter of choice, but prescribed (forced) by the underlying interaction of the P-laws. At the same time these P-laws limit, to a probability of zero, imagined – but ultimately non-existent properties. We take the properties of water for granted because they are so well known to us. Potentialities may be only known to us as the necessary outcomes of theories and hypotheses. These suppositions are not facts. Only expressed actualities, should be called facts.

So, what are these facts? The qualities of water that are especially notable to our senses are those of water in the liquid and solid form. The name, universal solvent, refers to water's ability to dissolve inorganic, ionic chemicals as well as smaller organic compounds. This makes it the "elixir of life." In the liquid form this "magnetism" attracts one side of the H2O molecule, the "O" side, to the "H" sides of two other water molecules. We see this in the phenomenon of surface tension — another emergent property. It is why liquid water, in small amounts, will form rounded hemispheres on a glass surface. Compare this to the relatively nonpolar ethanol (a two-carbon alcohol) which spreads out as a very thin-film. (Ethanol also has a small amount of polarity, but it is a better solvent for organic compounds for its own emergent reasons.) Surface tension is why a cup can be filled with water to slightly above the cup's theoretical volume. To a significant degree the water, that is above the rim, is held in place against gravity by the polar forces. This fine balance, between the forces of polarity and gravity, can be broken by the addition of a single molecule of water causing the excess volume to spill down the side of the cup. This phenomenon, of a single, seemingly insignificant change, has been given the name of "tipping point"— a referral to the instability of a perfectly balanced see-saw. A tipping point is another emergent property, namely a discontinuity in behavior, as one set of forces and its properties are suddenly overcome by another set. This can be shown mathematically in simple systems. Much of otherwise inexplicable, human behavior can be understood as the result of such tipping points in the activities of the brain- mind. There it is actualized in the all or none firing of a neuron as described in § 7.3.

We now turn to water in its patterned solid forms: ice, snow, and frost (higher degrees of randomness lead to configurations such as sleet and hail). The allowed crystal patterns of the former forms are also a result of polarity. Ice is a crystal structure reminiscent of those seen in heavier minerals. Ice, as found in nature, is technically a mineral. Snow, in its six-sided configuration, demonstrates some limits to the allowed patterned structures. The tremendous and wondrous variations based on the six-sided theme is a result of initial irregularities of the central core, variations in temperature, moisture, speed of fall and so on. (For extended examples see geology.com). Here we have multiple versions with a vengeance. Each separate microscopic environment, with its multitude of variables, brings out new forms — but in limited structural examples.

Emergence is neither chaotic nor random. It is not magical or otherworldly. It is the P-laws expressed in a complex environment. Perfect chaos, that is a universe without rules or structure, does not exist in nature any more than a perfect Platonic circle. Both are mental models that describe a level of absolute perfection, which a universe of three-dimensional space and omnipresent variable energy fields cannot attain. A circle made of matter has unavoidable miniscule variations at the edges at the atomic and subatomic levels of measurement. When we smooth out the differences by the use of equations, we get the ideal form – from which the idea of perfection arises. Reality does not conform to our ideal imagined ideas. However, very close approximations to perfection, approaching that of the axioms of mathematics, allows for useful manipulation of our environment.

Chaos would be a negation of the second principle, which sets the limits of the possible. But the ideal of randomness is also not attainable. Ideal randomness implies change can come about without a cause. Causes are all, at base, transfers of energy. Pure randomness implies a total lack of cause, namely a total un-relatedness of all things, which is a universe without the first principle. Therefore, the word, random, should actually be, and is technically used, only to refer to our inability to describe or predict the specific outcomes of complex causal chains in the transfers of energy.

Another variety of solid water, frost, as seen on a window (the old-fashioned single-paned kind), shows a different pattern than ice or snow, and

Matter/Energy, the P-Laws and Emergence 25

is often described as fern-like. Who has not been amazed by the sight of these fern-like patterns seen on such a frosted pane? This is illustrative of the limits available, and forced, by molecular properties. Similar patterns are forced on both inanimate frozen water and a living thing, the veins of a plant. They are also found in the veins of a butterfly wing and many other living examples. This pattern is a common example of how all things are related through the potentialities and properties allowed in the interaction of molecules. Emergent properties are not arbitrary and not limitless.

Ice also floats in its liquid cousin, water, because polarity causes the same number of water molecules in ice to occupy more space than in its liquid form. That is, it has a lesser density. In fact, the density of solids is commonly compared to that of water in its liquid form at 4°C. This ratio is called specific gravity, another example of the centrality of water to our understanding of the world.

Emergent properties are all the properties of known entities actualized in this universe. It has the limits imposed by principle #3. It should be made clear that this principle does not deny the existence of other universes or other transcendent realms that include deities. Rather it denies the possible interactions of our universe with the proposed and imagined extra universes. My understanding is that such interactions are not allowed in the equations that propose such alternate universes. At the same time, it implies that if any such interactions should occur, then it would result in a merger of the universes. If the P-laws of another universe are different from ours, and these systems merge or interact, then they both would perforce become part of a new one. This new universe would then have different, conflicting emergent properties. In this contradiction, the outcome would not be the unpredictability of chaos, but unpredictability never-the-less. Such randomness is not imposed by lack of information and time, but by the incompatibilities of causes. Such combined universes are incomprehensible, and therefore not helpful in a search for wisdom.

Principle #5 emphasizes the concepts of relationships and interactions. Interactions in turn imply change in physical structure and/or energy content or its form. Without change there would be no potentialities, emergent or otherwise. Change furthermore involves time. This is explicit in the famous equation $E=MC^2$ of Einstein concerning the relationship of energy and matter. The "c" is the speed of light as measured by distance divided by time! Finally, because of emergence, predictions of the properties of parts, when only the whole is perceived, are fraught with expected uncertainty. Emergence also explains why the properties proposed in quantum physics are different, and submerged, when the properties of the next level of matter, bosons (electrons, protons etc.) come to the fore.

Together, the five principles are the basis on which atoms and wisdom can be related, albeit through many, many levels of emergent properties. Subsets of the relationships in space, time, and energy can be categorized as systems incorporating different levels of emergence. The systems, with their various levels of possibilities, take us from the inorganic realm, to the organic, to life, to organism, and finally to mind. And it is one of the activities of mind that creates philosophy and the search for wisdom. To better understand what we mean by mind, we must explore the ontology of mind, one of the subjects of Chapter II. But we start that chapter with an expanded exploration of emergence

Chapter II
Emergence of Life and Mind

2.1 How Emergence and Ontology Allows Life & Mind
The emergent properties and potentialities of water were described in the previous chapter. But emergent properties already arise in the deepest subatomic reality. Each level of organization thereafter allows for new properties and submerges some of the properties of its constituents. If this were not so, our universe would be without the many distinguishing parts as described in "The Big Picture" Carroll (2016). The emergent potentialities of most interest to philosophy are those that led to the possibility of the activity called mind. To show how mind is connected to the rest of reality, we take a short tour of the levels of emergence. These levels arise by the increasingly complex organization of matter and the resulting energy flows. The diversity made possible by the various atoms, via these newly available properties, increases exponentially as they organize into molecules. Molecules can then further interact to lawfully create larger structures and thereby create new entities as described below.

Working backwards, we note that the activity of mind presupposes a complex system such as our brain. Our nerves, and the other structures that compose our brain, do not operate independently from the rest of the body. The brain is necessarily a part of an organism composed of specialized organs that supply, among other things, the energy needs of the brain. Organs, in their turn, are composed mostly of individual cells, the smallest living unit. The cell is where the property called life exists. The many functions of a cell, that together allow life, include metabolism, repair and reproduction (See § 3.3). Metabolism is the set of intracellular chemical processes that lead to the transformation of energy and builds and destroys structures needed for the function of each living cell. None of these transformations, individually, have the property called life. These processes are the result of the emergent possibilities of many specific molecules, themselves made up of atoms. The same relationships obtain for repair and reproduction.

The list of the major physical structural stages that eventually allow the newly emergent properties of life and then mind includes: subatomic entities, atoms, molecules, organelles, unicellular organisms, undifferentiated clumps of individual cells, specialized cells (such as neurons) in multi-organ lifeforms, and the specialized organs such as muscle, liver, and brain. This eventually led to the activity of the brain called mind. The details along the way are, of course, enormously complicated, but the underlying emergent potentialities can be stated more generally and therefore more simply. To reiterate, the increase in organization at each level is of paramount importance. For example, picture a vat of all the specialized cells, derived from a complete complex organism, but now living as free-floating cells. It is no longer the organism. It contains all the materials, but doesn't have the same organization and therefore not the same emergent properties. Furthermore, as structured combinations these living cells form a more efficient local environment where the component cells can thrive. In the vat, although all the individual cells are there, they do not constitute, without the organization, an insect, dog, or human being. It follows that "the self" arises via the organization, not only its parts. I am made of living cells but have other properties as well because of emergence inherent in the organization provided by my DNA. For a more extended discussion of how the vat "mental experiment" helps in understanding personhood (See § 10.6). An organized whole is more than the parts because of the emergent properties.

2.2 Brain-Mind

Brain-mind is not a property of life per se, but rather the property that is beyond the possibilities of a multi-cellular organism without a centralized nervous system. Its ontology is further developed in Chapter IV.

The history of the concept of mind parallels, and has been dependent on, the accumulation of information concerning human thoughts and actions. The brain-mind is not an entity made out of cells or a transcendent spirit. It is a process. It is best understood by comparing it, metaphorically, to metabolism.

Emergence of Life and Mind

Mind is the result of activity of a complex nervous system composed of living cells. These actions are those resulting from the pattern of communication between individual cells and systems of neurons. A non-organic basis for a mind is a possibility, but far from current artificial intelligence attempts.

What does the process mind include? At this point in our understanding a partial list includes: abstraction, imagination, emotions, logic, modeling, evaluation, conscious initiation of muscle activity, and its unconscious coordination between groups of muscles. Others can be added, and of course some work together or in a recursive tandem. This will be fleshed out in Parts III and IV.

In a universe such as ours, and if the P-laws are correct, such complex activity is the only type of process that makes mind possible. Minds based on different P-laws would lead to other properties, but similar functions. Such possibilities have been imagined, but without the coherence and consilience provided by the P-laws and the science-based models described above. It is the reason that I differentiate brain-mind from "artificial intelligences," processes which at best can only mimic a few of those of a living brain. The billions of parallel, recursive and idiosyncratic nerve impulses are beyond the reach of models less complex than our system of living cells.

Then, the first question that needs to be answered is: what are the necessary and sufficient components of a nervous system which allows the emergent properties that differentiate a typical human brain mind from those of simpler nervous systems? Comparisons between electro-mechanical computers and such simpler organic systems might be appropriate, but still fraught.

To answer this question, we must start with the basic unit of life, the cell. The cell is not just a bowl of soup, that is water and a bunch of chemicals floating around. The emergent property that we call life, although based on an accumulation and organization of water and chemicals, cannot be described (or occur) in one big jump. A short summary of the emergence of underlying properties will be necessary to explain why this is so.

2.3 Structure and Organization

Structure is the word for the spacial and temporal arrangement of things relative to each other. For material things the structure is in the dimensions of space. For intangibles, such as social relationships or positions within an organization, a spatial metaphor is often used when models are needed to describe and define these ordered relational patterns. Structure is necessary but not sufficient for organization. Organization implies a pattern that allows a particular set of results. In the inorganic world a crystal is an excellent example of a structure that leads to a specific set of properties which an unorganized collection of the same compounds cannot have. Structures remain organized when they are stable, usually in relatively low energy environments. An ice crystal will not change its structure if there is either no change in temperature or no physical force is applied that is stronger than the chemical bonds holding the atoms in place. An organized structure is one of the requisites necessary for a cell to have the properties of life. This leads to a pertinent anecdote.

2.4 Life on the Sun – A Telling Misconception

A few years ago, in a discussion with me, a freshly minted faculty member, with a PhD in philosophy, claimed that "life is possible on our star, the sun". I retorted that it was not possible because of the vast amounts of energies involved. He answered, "Well, that can't be true because there is life around hot ocean thermal vents." The fact that the temperatures around hydrothermal vents are in the range of 60 to 460 degrees Celsius, and the temperatures in and around the sun are in the range of 5000 Celsius at the surface to 15 million Celsius at the core, did not impress him. He had no idea of what the physical properties of matter are at such temperatures, nor did he know that the stabilities between nuclei and electrons are limited to exceedingly short periods of time at these energy levels. Most depressing (to me) was the fact that he did not seem to feel that the P-laws of nature had anything to do with how "life" was to be defined. He seemed to want to have the properties of "life", without the underpinning properties of the things of which earthly life is composed. In other words, his philosophy resembled fantasy fiction rather than an endeavor based on reality. I don't blame him for this misconception! The moral of this short vignette is not that there are philosophers who do not base their philosophy on demonstrable reality, but that a young and intelligent man came from an educational system that did not give him the most basic training on how to evaluate reality.

Unfortunately, a lot of what has been written concerning philosophical ideas, even in our age, shows the same disregard for the basic P-laws and emergence that is well documented. A short review of the kind of properties that emerge, and the necessities and limitations of our reality, is a core purpose of this book. There have been many scientifically-oriented philosophers such

Emergence of Life and Mind 31

as Bertrand Russell, Mario Bunge, Hunter Mead, Patricia Churchland, E.O. Wilson, Nicolai Hartmann, and many others, whose ideas, directly or indirectly, added to the general understanding of our world. I have tried to summarize, integrate, promulgate and possibly add, in a small way, to this understanding by writing this book. It is hoped that it will add to the "The Great Conversation" of Norman Melchert (2014).

2.5 Beyond water and inorganic chemistry —Organic chemistry

The emergent properties of water, the main ingredient of all earthly living things, have already been discussed as the polar qualities of its liquid form are of great importance. The next most common component for living organisms is carbon. Organic chemistry is the chemistry based on the carbon atom. The carbon atom has an electron structure allowing it to interact with up to four other atoms. This property allows the formation of a vast number of potential molecules. This statement is not hyperbole. It is based on the simple fact, that this element, written as "C," can form long chains, with various side branches as well as loops and rings in three dimensions. If one looks at the structure of the basic proteins (as depicted in a source such as Wikipedia), the complexity and possibilities will quickly become evident. A Wikipedia graphic can be found at:

https://upload.wikimedia.org/wikipedia/commons/6/6e/Proteinviews-1tim.png.

Carbon is an active atom, combining with not just itself and oxygen and hydrogen, but also with nitrogen, sulfur, chlorine and phosphorus as well. Furthermore, the complex structures that can be made out of these elements can be configured so that they can act as "containers" for metal ions such as iron in mammalian blood, copper in the hemocyanin of arthropod "blood", and magnesium in chlorophyll. Many other combinations form enzymes and structural molecules. The actual details are not important for philosophy, but the positive possibilities and the negative limits of atomic structure must be understood.

2.6 Why Not Silicon Instead of Carbon?

Directly below carbon in the periodic table of elements (a categorization of the elements based on their proton components and their chemical activity) is silicon, and therefore shares some of the chemical properties. The possibility of life forms based on silicon – replacing carbon — have been both used in science fiction, and considered in science. If even possible, the potentialities of such a life form would be very different because silicon is physically significantly larger than carbon. This changes the geometry of its possible compounds and also of their other physical properties. For example, carbon dioxide at the pressures and temperatures on the surface of our earth is a gas, while silicon dioxide is a solid. It is the major component of sand and glass. It can, however be found in carbon-based life as a structural component, its crystalline structure conferring durability to surfaces and supporting structures. It does not, however, have the flexible qualities of carbon. Therefore, the possibilities of life based on silicon, replacing carbon, are clearly very limited, if not impossible. It seems that life needs to be carbon based. To further investigate that idea, we will delve a little further into the necessary components and circumstances that allow life to form and persist.

One of the properties contributing to the success of carbon-based life is the structural folding that is possible. The strings and loops can and do fold in the various ways allowed by the carbon linkages. These folds in three-dimensional space are often only semi-stable, depending on the environment. This allows for environmentally sensitive feedback and control. The folds are not held in place by chemical bonds, but by the weaker polar forces that play such an important role in the structure of water. This type of flexible structure provides for the "lock and key" mechanism for many enzymes. The lock in this case is the enzyme in its inactive state; the key is a molecule that fits into the structure of the enzyme in such a way that it activates the protein to perform a task. This dynamic complexity increases the possibilities of chemical interactions exponentially. Basically, when two separate molecules come together and one is the protein, the protein can deform to the point that the probability of a specific outcome of the interaction becomes closer to (1). This is how enzymes work. When the products of the interaction separate, the enzyme snaps back to its original shape and is ready for the next interaction. This mechanism makes an enzyme a catalyst. It is estimated that the rate of such reactions can be increased by a factor of up to a billion. A more complete discussion of the importance of catalysts is given in a discussion of causation. (§ 4.6)

2.7 A Closed Environment--Bubbles

Before there was a living cell, there was the proverbial primordial chemical soup. One of the major and necessary structural components of a cell that differentiates it from this soup is the cell membrane. Membranes, like those that form a soap bubble (an enclosed space), can most certainly be formed de novo in such a soup. Of importance for the development of life are the characteristics of that membrane and the contents of what is now inside the membrane. Although we have little detailed understanding of these early processes, we do know that the dynamics of a closed system can be very different from that of one that is open to the general environment. The membrane of such a bubble can be compared to the walls of a test tube, albeit with some differences. For example, such a membrane, to be useful for metabolism, cannot be inert like glass, but must interact in specific ways with both the chemicals that are outside and inside a contained system. This kind of membrane is called semi-permeable, and can have both passive and active characteristics. The importance of such a closed, active system is that it changes the possibilities and the probabilities of interactions by virtue of the differences in the dynamics of an open system versus an enclosed one. In short, it allows a greater local accumulation of basic molecules such as alcohols, ethers, amino acids, sugars, fatty acids, peptides, nucleic acids, and so forth. Of these, the fatty acids form a major component of the future cell membrane. All these chemicals are the necessary precursors in the formation of the next level of complexity, namely organelles and other structures. This was a necessary step in the progress toward life, as discussed below. Exactly in what order these components came together is unknowable and perhaps not critical — but other necessities must also be met. In any case, the building up of various chemical components occurred nearly simultaneously – when the time frame is the tens of thousands of years.

2.8 Energy

Of major importance are the sources of, and mechanisms that capture, the energy that drives the chemical processes. The source, type and amount of energy must be such that it does not cause the destruction of the necessary building blocks. Recall the above parable concerning "life on the sun." Then as now, there are two main sources of appropriate energy available in this range. The first is thermal (heat) energy, one of which is the molten core of the earth, available especially near the venting of hot gases into bodies of water. The second is the radiant energy supplied by the sun, which in turn can be converted into heat and other forms of energy. The application of the scientific method to this problem has shown us what the characteristic limits of energy, useful for living cells, must be.

The earth-based source was initially probably more important, nonspecific and general. By increasing molecular motion, heat increases the rate of chemical interactions and thereby further increases their probability of the occurrence and the quantity of the resulting products. Some of the sun's radiation also contributes to the ambient temperature, but it only later became the major source of energy which allowed the continuity of certain life forms. The most successful and usable vector for this later phenomenon is the chlorophyll of plants. The importance of the flow of energy for the process we call life is stressed by Lane (2015).

2.9 The Conservation of Structure: From Nano- to Micro-Structures to Convergent Evolution

The pathway to the chlorophyll structure was obviously not made in a single step, nor in a single century. Its arrival on the scene depended on the availability of the building blocks needed for its creation and later the genetic changes to make the catalytic proteins for this new reaction. The sciences of biochemistry and biology have determined what these building blocks can be. The information regarding the genes responsible for the creation of the responsible enzymes is now being slowly but steadily elucidated. It is of some interest that there is a similarity between the metal containing structures of chlorophyll and hemoglobin. The main difference is the metal component: magnesium in chlorophyll and iron in hemoglobin. This points to a fact seen many times in nature, namely, that the same building blocks are used to make products with different potentialities. An example of similarities in development (at the next level of structure), is found in a comparison of the "veins" of the leaves of some plants and those in the wings of a butterfly.

This points to the fact that some structural forms are more successful than others. It may seem remarkable how similar the pattern is — but it should not surprise us. The P-laws determine what kind of structures facilitate the even distribution of fluids over a planar surface or through a tubular structure. A few patterns have similar potential and efficiency and we see them again and again.

At the more complex level of organisms, we see the occurrence of convergent evolution. This describes the phenomenon where very similar macrostructures have their source in a completely different genetic background. Wings are seen in insects, birds and bats. Sharp teeth are found in carnivorous fish and mammals. Photoreceptors, including eyes, are very common in all the levels of evolution.

However, the primary and necessary characteristic of evolutionary development is not the change, but the retention, of systems (like DNA and RNA) that are successful in keeping an organism as a link in a chain of life. This fact, namely the great conservation of DNA structure, is the norm. Mutated genes, as proportion of all the copies made, need be exceedingly small in number. The macro patterns that are based on the microstructure are too specific in their functions to allow more than minimal change in the genes in any one generation.

A good example of this, in two ways, is hemoglobin, the structure responsible for the transport of oxygen in mammalian species, among others. There are sub-types of hemoglobin in the animal kingdom and further sub-types in humans. In the latter, hemoglobin A (the norm), S, C, and E (and hemoglobin F in fetuses and newborns) are the main variants. In this case the change is in one of the two protein chains: the ß-chain, that make up hemoglobin. Each variant is most commonly due to only one change in the one hundred and forty-six (146) amino acids that make up a ß-chain. These variants cause different degrees of anemia (low number of red cells) but are partially protective against the malarial parasite that attacks the red cells of blood. Importantly, evolution is not a response to the parasite, but rather the improved viability of the already formed variant. Hence the phrase "survival of the fittest." If there are changes in the α-chain (one hundred and forty-one (141) amino acids)) the results are called thalassemias, another cause of anemia. It is possible to have a combination of the two types and there are also genetic changes that do not cause any noticeable differences in oxygen carrying capability or other pathologies. None of these is to be confused with blood types (A, B, O) — though similar comparisons can be made there.

The second most common is hemoglobin F, which works better in the lower oxygen environment of a fetus. This variant also is formed by a difference of one amino acid out of the two hundred eighty-seven (287) that make up the two component proteins that form the molecular "cage" for iron. Most other changes usually affect the molecular structure to such an extent that the hemoglobin can no longer carry out its function. The same is true of most other structural proteins and enzymes. There is variability, but usually only minimal changes to a gene at any one time, will allow survivability to another generation. This again emphasizes the fact that, although the differences are the most obvious, the similarities are the most critical.

2.10 Possibility, Probability, Property, and Randomness

The possibilities (potentials) of an event in a new system are a direct consequence of emergence. One commonly speaks of the properties of a system, but a property is better understood as an actualized possibility. There is also a difference between possibilities having been actualized versus demonstrated. A thing may be real and have material existence without having been demonstrated by science or otherwise. The reality of an energy type, physical entity, or a relationship between them, may be inferred in such a case from their place in a theory such as the proposed existence of the Higgs boson in the 1960s. This subatomic particle (and field), or rather its properties, were postulated as a result of equations based on theories in particle physics. Said another way, it is required, if a particular model of reality is correct. Experiments in the second decade of the 21st century seem to have confirmed its actuality. This was after the Large Hadron Collider was built at the European Organization for Nuclear Research (CERN). By the use of this very complex (and very expensive) instrument of science new measurements were a match for those expected from theory. As of this writing, some experiments match 99.9%, but others need more measurements to get more accurate results. Why has this taken so long? Short answer: The demonstration of an actualization of a possibility depends on the probability of our ability to observe its effect. In this case only by the use of the CERN reactor does this possibility reach the level of expected observability.

A less esoteric example can be imagined. A priori it is possible for an infant girl born in China on April 19, 2015, to someday meet a boy born, on the same day, in the Amazon Basin of Brazil. If they do indeed meet, that would become a property of their relationship. The probability is of course exceedingly low, given no other connections than that they are human and live at the same time. The possibility nevertheless exists, whereas there is no possibility, a probability of (0), if they were to have been born on the same day, but two hundred years apart. A property such as a personal relationship is a potential interaction that can only be confirmed after-the-fact.

Randomness is also often misunderstood by those who feel randomness indicates a lack of causality. An actual meeting of the two above mentioned children, when not positively arranged by another agent, would typically be called a random occurrence. This, of course, does not mean that there were not thousands of small causes, which in conjunction, actualized the meeting. What random means, in such a case, is that that meeting — and its probability – could not be calculated, or predicted as an actualization. The very small possibility of a meeting remains until one of them dies. It is most probable that they will not meet. But if they do, it is because of many necessary but undeterminable causal events (false randomness). The causes may be, even in retrospect, indeterminable, but the possibility existed all along. Randomness is no more than a statement of uncalculatable probabilities. And so it was with the formation of earthly life. Life exists, and therefore was not only possible, but by its very existence its probability is (1), its formation has occurred. This is so, even though all the causes have not or cannot be demonstrated. No matter how small (other than zero) the probability of any of the causes that led to life were, they occurred and had their effect. When the acting agents are not clear to us, our ignorance does not affect the probabilities for the occurrence of actualities. Arguing that something cannot be because we don't understand it is illogical. Positing the impossibility of a potentiality should only be based on the logical extension of the P-laws and demonstrated facts. There is, however, an important difference between proposing a necessary part of a theory, and a fantasy. Predetermined fate is not demonstrable nor deniable and it is usually ascribed to transcendent teleological forces. These differences will be shown to be very important when we speak of knowledge in part II of this book.

Probabilities are themselves emergent properties of the system and are often misunderstood, misinterpreted, and misused including in some scientific endeavors. This unnecessary confusion is what leads to the feeling that an Aristotelian "final cause" is necessary for some descriptions of an action in our universe. A short description of Aristotelian causes and a modern version is given in the" Vocabulary and semantics in philosophy" § 4.5 below. For the time being, it is enough to indicate that a final cause presupposes an agent that has a goal.

2.11 A Short Excursion into Mathematics

Probabilities are expressed as a number between one and zero (1) & (0). Anything that actually exists, or has existed, has or had a probability of (1). Only something that is impossible (based on the actual P-laws of our universe) has a probability of (0). Probability has no meaning in transcendental world views because they cannot demonstrate immutable laws. In a world of P-laws, potentialities, anything that is possible, have some probability greater than zero of being actualized somewhere, at some time. The minimal probability can be expressed either as a fraction: $(1/Y)$, where Y is the number of all possibilities — or decimal: $0.X$, where X is any positive string of numbers that is not zero. The greater the number of emergent potentialities, from any starting point, the smaller a priory probability of each actualizing into a property. This is because all the probabilities must add up to (1). Any difference in the local inputs (set of properties) alters the probability of the individual results. For example: only a perfectly cubic die with its center of gravity at the exact center will have the potential of each face coming up an equal number of times in an infinite series of throws. All typical dice will come close, so close that it does not affect the expected results in a realistic number of tries. Furthermore, the average of a large number of "imperfect," but not "loaded" dice will approach the results of the ideal one due to the averaging of potentialities and therefore the results.

At this point it will be helpful to grasp the immensity of sub-infinite, but very large, numbers by expressing them as exponents of 10 (such as $10^2 = 10$ squared or 100; $10^6 = 10$ to the sixth power or one million and so on). The number of atoms in the universe is currently estimated to be between 10^{79} to 10^{81} — not counting any that may exist as inferred black matter. The number of positions on a chessboard, the Shannon number, is on the order of 10^{42} to 10^{50} depending on the definitions of allowed positions. These very large numbers of "Y," and the even larger possible spatial, temporal, and energy relationships, make any one future for each atom incalculable, that is seemingly random. But improbable conditions and results not only exist, but under certain conditions will actually be common locally.

A simple thought experiment based on atoms makes this clear. If the number of atoms of a rare element Q has a ratio of $1/10^9$ (one in a trillion) of all atoms, and if there are 10^{80} atoms in this universe, then there will be $10^{80}/10^9$, that is 10^{71} atoms of Q in entirety! Scattered all over the universe this element would be hard to find. However local collections of an element occur because of local conditions; in this case stars that were pulled into existence by gravity. This led to the emergence of varying, but new, probabilities of new atoms being formed by fusion and fission in the core of stars. Gravity is a variety of energy the properties of which are only partly understood, but it is a necessary force for our universe to exist as it does. In

Emergence of Life and Mind 39

short, local conditions such as those in a primordial star, which is a very dense collection of hydrogen and helium nuclei (and eventually heavier atoms in secondary stars) increase the probability of possible nuclear interactions and thereby local concentrations of the heavier elements. Suns do not have membranes to hold them together, but gravity serves the purpose of concentrating matter in an interactive cluster.

In summary, low probabilities lead to low actualities which then leads to difficulty in observation but this does not equal impossibility. Impossibility necessarily implies a probability of (0). These concepts will be shown to be also important in our discussion of knowledge (Part II).

2.12 Probabilities in "Bubbles" Revisited

Local concentrations of matter increase the probability of some of the possible interactions. In chemistry, when the scale is sub-planetary, this localization can be increased by physical boundaries, a container, which limits the dispersion of the molecules. Containers, such as flasks and test tubes, are seen every day in the laboratory. In nature such containers also abound. Depressions, holes both large and small, in rock and soil are natural containers. Evaporation of water in such a container would further concentrate anything dissolved in it, bringing the solutes closer together, again increasing the probability of some of the possibilities. But too much evaporation would decrease the mobility of the solutes and thereby decrease the chance of interaction. However, that state does prolong the duration of contact, which again affects the outcome.

These kinds of conditions change, and sometimes increase, the probability of specific chemical interactions between the components that are in such a system. This allows new molecules to accumulate, which then changes the potentialities and probabilities of further reactions by changing the new initial conditions. In very general terms, the possibility of life forming depended on the accumulation of chemical structures in such an enclosed environment. This process of accumulation of new molecules led to the further emergent chemical and structural changes necessary for the next stage in the development toward the increase probability of a living cell. This illustrates how the relationships and differences between possibility and probability are important, both to science and philosophy, or to any other discussion of reality. Our level of understanding of the possible refers to the conformation of the model, to a suggested real thing or relationship, and to the underlying P-laws of our universe. By extension, this conforming to those P-laws includes those relationships that emerge from more and more complex structures and organizations of matter. Being possible means that given a certain specified state has the probability of an actualization of a specific outcome that is not (0). Nevertheless, sheer possibility does not imply necessity, that is a probability of (1). What structures such as surfaces, bubbles, and membranes provide is the emergence of possibilities and increases in their probability. The limits of possibility can be stated as either positive, greater than zero, or negative, zero. The probabilities, however, can change by factors of millions by the action of catalysts, as first noted in § 2.6. Without those great increases in probability in the formation and accumulation of basic chemical structures, the probability of life arising would be very near zero. But each step of organization and complexity increased the probability and, at least on our earth, reached the probability of (1). That life exists is not controversial, although what its properties are may be a matter of discussion. The causes and mechanisms leading to a significant probability of each of the steps leading to life is one of the disputes between scientific and transcendental thinkers.

2.13 Knowledge as a Tool

Until recently (a few hundred years ago) science has not had sufficient physical tools to demonstrate the possibilities, and the factors, leading to the increased probability of the formation of life. The following thumbnail sketch of the steps leading to multicellular organisms and nervous systems is provided mainly as a source that points to the contents of the "book of knowledge." This virtual book contains current scientific data, interpretation, and understanding. This outline is also meant to be a source of a basic vocabulary that can lead anyone with access to the internet, or a good university library, to an examination of this extensive pool of information. The following section is provided with that purpose in mind, but it cannot provide all the detail necessary to form a highly-informed personal opinion. Rather, it is an attempt to show why the scientific endeavor slowly and progressively leads to opinions (interpretations) that are based on the consilience of the confirmed data. These models are what is commonly called and comprises knowledge. Science provides a strong infrastructure of detail, all interconnected, that imagination alone cannot provide. That is how progress in philosophy will also occur. Philosophy deals, especially, with human interactions, which are near the end of one of the emergent pathways on earth. Ignoring the demonstrated pathways will, however, lead not to progress and increased wisdom, but to confusion, intellectual dissonance and needless argument. Again, wignorance is not a useful option. For this purpose, chapter III on basic biology is provided. It is meant to provide a consilient basis for a philosophy based on demonstrated and deduced reality. Before we proceed, however, a few words need to be said about the relationship between actual reality (or the-thing-in-itself also called Noumena by Kant) and mental models.

2.14 Reality, And Its Models – A Preview of Epistemology

That there is an actual reality that does not depend on the existence of humanity or any sentient being could be seen as the unwritten Axiom #0. The existence of some reality underpins all the other principles; it is their referent. For purposes of discussion the mental models of this reality can be categorized into three types.

These models are formed, by the imagination, by two main processes. Deductions are based on all the information that is consilient with the basic five principles stated above (sec 1.3). Inductions and abductions are less definitive. These inductions of reality are created when there is a lack of consilience with the basic principles between current data and/or a lack of information concerning the relationships. All mental models use the imagination to fill in the gaps, as more fully discussed in Part II. As such, the induced models lie between deduced models and fantasy. Fantasy is a model created, as all the others, in the imagination, but the term indicates that aspects of this kind of model are contradicted by the basic principles or the deduced models. A change in available information can transform a model from one category to another, and of course a different set of principles would change the classification of the models. The types of models become important in the forthcoming discussion in Part II of what counts as knowledge concerning the Noumena.

Chapter III
Possible pathways to life

3.1 Connecting Pre-Biologic Systems and Life

Because of a lack of detailed information, the imagined but possible pathways leading to life belong in the realm of the inductive models. There can be differences in assigning probabilities to these proposals, but they are not zero. With more information now available, they can be provisionally ranked. With time, sub-models have and will rise into the deductive model range. The main purpose of proposing these models is to show that models which belong in the fantasy realm are not needed for a philosophy.

3.2 Repair and Maintenance – The Importance of Further Organization

The next types of organization, with its attendant emergent properties, are made possible by concentrated chemical environments. These kinds of local conditions were involved in starting relatively simple feedback loops. In a way, this can be seen as the first signs of a system we now call evolution. It is, after all, those systems that multiplied themselves that came to dominate.

In the simplest cells, bacteria, we see that a major component of the repair and maintenance processes are cellular structures that are not membrane bound. These are the physical elements which, by their very structure and components, increase the rate of certain chemical reactions by orders of magnitude, not just by additive progression. That is, catalysis occurs. It can be surmised that one of the most important reactions would be the creation of chemicals that keep the cell membrane (the boundary of the "bubble"), intact against damage from the environment. After that, myriads of permutations and combinations of those chemicals and structures gave rise to containers that could keep themselves stable for longer periods of time. This then allowed further developments that eventually led to what we would recognize as a cell. With the modern tools of science, which allows investigators to peer into that structural level, it is clear that a cell is much more than a bag containing a soup of chemicals, however complex. Instead, the cell is composed of a basic set of structures laid out in planes, tubules, vacuoles (microscopic bubbles) and subcellular membranes. Water, electrolytes (salts, especially sodium, potassium, chloride, calcium and carbonates) and small organic molecules such as sugars can move easily in the spaces in between. But many important reactions occur on structured components including the organelles where many of the enzymes are found.

3.3 Defining "Life"

At this point a definition — that is a description of the characteristics — of life will be quoted before we proceed.

> "*Life – macromolecular, hierarchically organized and characterized by replication, metabolic turnover, and exquisite regulation of energy flow, - constitutes a spreading center of order in a less organized universe.*" (Grobstein, C. "The Strategy of Life" (1965).

From the discussion above, we can see that I basically agree with Grobstein, but with the addition of the repair mechanisms without which life cannot persist very long. It also seems clear that the organization needs to be within a bounded structure, what we call a cell. The definition then would become:

> "*Life — <u>bounded</u>, macromolecular, hierarchically organized and characterized by replication, metabolic turnover, <u>self-repair</u>, and exquisite regulation of energy flow, - constitutes a spreading center of order in a less organized universe.*"

A caveat is that replication is only necessary for the continuation of a specific life form, but not for its presence at one particular point in time.

At this point in time, our search for understanding concerning the beginnings of life lacks the precise information needed to write a clear or precise history of this beginning. We do however have enough information about life as it exists to hypothesize and test the steps necessary to achieve this level of organization. However, even if we manage at some point to create a cell out of components and building blocks which created in a test tube, this would only show that that pathway is one possibility. The actual sequence of events can only be imagined and will always be unknowable in its details. Creating a living cell de novo would only show that our basic understanding of the pathways from organic chemistry to biochemistry is on the right track.

For several reasons it is also very unlikely that the original sequence will repeat itself, at least on earth. First, once cells became common, they began to compete for resources which included the pre-cellular, chemically formed organic compounds. This made these compounds less available for a natural de novo start. Secondly, and perhaps more importantly, many other pertinent environmental factors changed with time. Has life, or pre-life, begun elsewhere in the universe? Until new information contradicts it, the answer must be, "Yes, it is possible". That places the probability to a value greater than zero, but does not guarantee a probability of one. Only finding non-terrestrial life would confirm the actuality of this possibility.

3.4 The Emergent Properties of Life

The advent of life led to another group of emergent properties. These can be conceptually organized based on the basic needs and organization of the living cell. The most basic emergent properties are those that are compatible with a living cell: maintenance of a complex equilibrium, repair, energy management, reproduction (clonal), and viable mutations. This list therefore excludes the red blood cell of the circulation because it lacks the necessary DNA. Each of these need-based general properties have been met by various cell lines using a variety of mechanisms.

For metabolism and mind to exist, life must exist. The word "metabolism" subsumes all the chemistry-based activity of a living cell. As we shall discuss in Part II, the word "mind" subsumes some of the activity of a nervous system. But to claim a relationship to reality, the details that define the properties of mind must be limited by the possible. If some other highly complex, non-living, structurally based mechanisms are available to create the process we call mind is an open question. Only then, perhaps, one can begin to talk of the possibility that such an entity, with this emergent activity, can philosophize.

On earth, after the success of the single celled organisms, the specialization of cells was the next step on the road toward the actuality of mind. Such cellular specialization increased the probability of complex life forms' continuance in several way. Among them are the possibilities created by symbiosis (the combination of specialization and cooperation), parasitism, niche utilization, and finally the flexibility that is the hallmark of evolution. (See § 13.7).

3.5 Cellular Differentiation

How do cells of the same multi-organ organism, which have the same chromosomal DNA, develop a different set of characteristics? The answers to this seem to lie in the local, internal and external, micro-environment of the cells. The complexities of the mechanisms that switch the activation of specific genes on and off are still only partially understood. Again, for philosophy, it is more important to understand that there are such verified, scientifically consilient mechanisms rather than what the details are, and that therefore, no transcendent forces need to be invoked to explain their emergence. A complete philosophy only demands adherence to the best information available. When it comes to mechanisms, science leads the way in such understanding, and its findings need to be consulted.

3.6 Growth and Early Reproduction

A natural progression of the above-described repair and maintenance mechanism would lead to growth, that is, to an increase in the number, size and complexity of the components necessary for stability. At some point a larger cell would become unstable. Recall a large soap bubble that can change into two smaller ones under the influences of the forces of gravity and mild breezes. The alternative is that the membrane breaks and the bubble ceases to exist. The new smaller bubbles are more stable. Such pre-life divisions would be the first step in what results in asexual reproduction. Asexual reproduction is, in its simplest form, nothing but growth followed by a nearly equal division into more stable smaller units. In a recurrent cycle these clones are the advent of a species.

All the details, such as which chemicals came first, which kind of membrane initiated the first stable structures, and of course the temporal progression of the changes is mostly conjecture. But by applying the knowledge of which minimal physical forms, relationships and chemical and physical interactions are necessary to maintain a living cell, in the various current environments, we can formulate the limits of the probable chain of events. At what point such a dividing bubble can be said to be a living thing is a somewhat arbitrary decision, but the amended Grobstein (1965) definition would seem a good yardstick. At what point the process became reliant on what we now know as genetic material can also not be precisely ascertained. However, this genetic material (initially probably RNA) and its inherent self-duplicating capabilities greatly increased the stability of the mechanisms needed for repair, maintenance, and growth. These nucleotides-based molecules are crucial because of their nearly unique self-replicating properties. (The growth of crystals and the multiplication of prions are other examples). The nucleotide-based compounds also have the ability to act as catalysts (See § 4.6). This ability they share with some proteins. And, the myriad types of proteins also play major roles in the many other characteristics of a cell.

The order of these further changes in the evolution of these very early cells (to which we now ascribe life) is again impossible to know in detail. We can now, however, examine the end results (species) which all seem to be related to the earliest forms at the deepest levels. That connectivity, beginning in the relationship between a parent and its daughter cells (arising by division), is what is meant and understood as the unity of the tree-of-life.

3.7 From RNA/DNA to Organelles

At the same time that various amino acids, sugars, and fatty molecules increased in concentration by pre-life processes, a fourth kind of compound, the nucleotides, also increased in number. Whereas the first three components are part of the physical basis of the cell, the latter formed nucleotide chains, essentially becoming pre-RNA and pre-DNA. The names of the nucleotides that formed these chains spontaneously and became prevalent, are the ones we know of today: adenine, cytosine, guanine, thymine and uracil. These chains have the unusual property of acting as templates which, by alternating steps, first form a complementarity molecule. Then this complement acts as a template for duplicating the original. Furthermore, some of the chains increased production of specific proteins. The point for philosophy is not how this process progressed, but rather that it is possible, and was actualized. Science concerns itself with the finding of possibilities, probabilities and actualities. Philosophy deals with what all this might mean and portend and what value it may have. An individual brain- mind can manipulate the models of mechanisms and meanings using the imaginative process. When combined with the results of the efforts of others we can all come to a more complete understanding of reality. This effort is the purview of Part II, Epistemology.

Currently scientists are vigorously studying the tips of the evolved branches of life. By applying the rules of physics, chemistry, and the emergent rules of biology, conclusions concerning the most likely paths to the present can be reached. Of the many branches of life, one is the human species. To describe and understand the species-specific possibilities and limitations we need to summarize a few more processes.

The next step in the evolution of cells, as noted above, would be the formation of organs made up of individual cells. Each distinct cellular structure is encoded from the same genetic material. The differentiation arises because only a subset of the commonly held genetic information is implemented.

Historically, between these early pre-life structures and the nucleated cells (eukaryocytes), out of which multicellular life-forms, including Homo sapiens (H. sapiens), are composed of, are the bacteria and archaea (both are prokaryotes). In eukaryocytes the DNA is largely concentrated in an internal bubble with its own membrane, called the nucleus, whereas in the prokaryotes the RNA and DNA tend to clump together in one area of the cell without its own membrane.

The exact relationships between the two unicellular, prokaryotic life forms are still under investigation, but it is currently accepted that the archaea are distinct genetically and metabolically. Whether these distinctions point to separate creations of life forms is not known, but several commonalities point to divergence from a primitive ancestor more than two billion years ago. In any case, the existence of the two types of prokaryotes gives us clues concerning the minimal necessary components of life. One of the things these two have in common is that they both have internal unique structures – the aforementioned organelles. The formation of these organelles was the next step of organization that increased the probability of success of the first life forms so long ago.

The length of time from only free-floating organic compounds to the prokaryotes is also estimated in the hundreds of millions of years. But the process was not linear. At each stage mentioned above, and at all the ones in between, the speed of change increased. The speed of change was somewhere between crawling and exponential at each point, depending on the inherent probabilities of the emergent possibilities as created by each new system of organization.

One other interaction that occurred over time needs to be mentioned for completeness. That is the incorporation of one type of cell by another. This is the process whereby prokaryotic electron transport chain structures attached to the plasma membrane became the mitochondria of eukaryotes. Similarly, the related chloroplasts (the organelles that harness solar radiation for useful energy) were incorporated into eukaryotic cells which became the precursors of plants. These acquired organelles increased the efficiency of the cells that came to contain them. Both organelles contribute their own genetic material the structure of which is similar to that of specific free-living prokaryotes. This has led to the conclusion that one type of cell, a nucleated eukaryote incorporated all or parts of a simpler prokaryote. This process is named endosymbiosis. An extended discussion of the importance of energy management can be found in the book, "The Vital Question", by Nick Lane.

3.8 Organ Formation and the Nervous System

We do not need to dwell on the speciation of bacteria and unicellular eukaryotes other than to point out that with more and more diversity the probability of multicellular structures increased as daughter cells adhered to each other. The earliest and simplest forms, of which we can still find modern examples of, are sheets, globules, and tubes.

Unicellular organisms developed not only internal organelles but also some structures that protruded through the cell membrane. They also released chemicals which became part of the environment for other cells, creating interaction at a distance (another emergent property). The advent of multicellular organisms allowed for further possibilities of interaction for these structures and chemicals. One of these possibilities, now well-documented, is that the chemicals released by one cell affects the function of adjoining cells of the same organism. Just as various molecules are arranged to form various structures and organelles within the cell, the interaction of the chemical products of one cell now affected the arrangement of structures in its neighbors. These effects, by which DNA is differentially activated in neighboring cells, is regulated by the changing chemical mix in the nucleotides' environment. This process increased the variability of structure and function in cells of the same organism. It is the basis of organs — the specialized cells within a multicellular, clonal life form. Eventually, one of these specialized cell groups formed the foundations of a nervous system. (See § 15.6 discussing clonal death). There are some very primitive species that illustrate the simplest function of organ development. The best studied perhaps are the algae. The variety within this group is in itself quite complex.

3.9 Sexual Reproduction in Eukaryocytes

After the differentiation of cells as noted above, another early change occurred when two types of cell division became possible. The first is the "old fashioned" kind of division. Cells grow, each RNA/DNA strand (chromatid) is duplicated, and then the chromatids separate by the process we call mitosis. This transformation into daughter cells (cellular clones) by cell division results in each cell having the original number and type of chromatids. It is the basis of all asexual reproduction. A second type of division subsequently arose – meiosis. Meiosis is the mechanism that makes sexual reproduction possible. This created a variety of chromosomal mixtures within a species that made subsequent speciation and evolution more probable. This process involves the creation of two haploid cells each with only one copy of a chromatid strand of DNA. The two kinds of haploid cells, the much larger egg and the tiny sperm, have the property that they can join to form a new, and unique cell with the normal DNA complement. This new process in biologic systems allowed the emergence of a superior method of species survival, as the new combinations of genes had the potential of responding differently to the same external environment. An increased variability within a species during reproduction now became inevitable. The pros and cons of the two types of reproduction are many, but it should be noted that very few multicellular organisms, and then only the simplest, can use either mitosis or meiosis for reproduction. Surely any beginning student of biology wished that the given names, meiosis and mitosis, were not so similar, with so much hanging on the differences.

3.10 Advantages of Multiple Organs

The physical organization of multicellular organisms dictates that different cells will experience slightly different environments. A simple tube is one such structure. If the flagella (tail-like protruding organelles) of the tube forming cells all point to the inside of the tube, and work in concert, the flow of the nutrient containing water past the cells can be increased. This increases the viability of such a structure over others that are dependent on the passive flow of water past them. This can be accomplished either by increasing the movement of the water past the cell's membrane or, by the use of flagella that move the cell through the water in an unorganized fashion.

These basic structures, globules, sheets, and tubules, are the beginning forms of all multicellular organisms, including H. sapiens. Embryology is a study of the growth, multiplication, reorganization, redifferentiation, and programmed death (called apoptosis) of various cell types. The basic processes controlling these changes are the same. They are regulated by the interaction of the DNA with its immediate intracellular and indirectly with the extracellular environment. Exactly what these interactions are, what kinds are possible, and what impact these interactions have on the development of specific functional proteins, is one of the exciting and fruitful areas of current biological research. In all cases however, when thoroughly studied, we see that the outcomes resulted from the emergent possibilities and probabilities. A multicellular lifeform's uniqueness is created by the relationships between the molecules, organelles, cells, and organs.

3.11 The Neuron — A Special Kind of Cell

The kinds of cells and structures that are of most interest to philosophy are the neurons and the simple and complex systems of which they are a part —the nervous system. The previous section of this chapter provided the minimal information necessary to show that neurons, and by extension the nervous system, are only a special case of general processes that led to life eons ago, and continue to support it. This section will discuss at greater length the emergent properties engendered by neurons in an organized nervous system.

Whereas the basic metabolism of neurons is the same as that of the other organs, there are several properties that distinguish them from other types of cells. What distinguishes a neuron are mainly two features. The first is its shape. Most cells are variants of the globular form as contorted by its apposition to other cells. Some cells with specific functions, such as muscle cells, are elongated; the white cells of blood float freely, mimicking the unicellular forms of amoebas. Liver cells are largely all the same, but arrayed in clusters. Other organs, like the kidney, derive their complexity from being composed of many different kinds of cells, all packed together in an organized manner, allowing the organ to function as an engine of cleansing and homeostasis (the maintenance of biochemical equilibrium). The same is true of the skin, which is a combination of cell types that differ locally according to the function of the surface covered. Neurons, however, have variations in the structure and the function of different parts of each cell membrane; these are much more extensive than any other cell type. Each neuron has four basic functional membrane types: the cell body which retains a more or less rounded form; the dendrites (the "receiving antennae" which are relatively short "branches" directly connected to the body); the axon (usually a single long branch which propagates a signal); and the specialized endings of the axons (telodendria), which form the proximate side of a synapse (the functional space between one neuron, and the next). The illustration below shows these, and some other features.

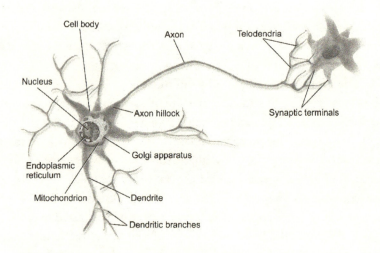

{Source: Wikipedia illustration under "Neuron"; by Bruce Blaus – Own work, CC BY 3.0}

Most of the neurons in the central nervous system interact with other central neurons. The others are of two major types, namely those that compose the afferent and efferent neurons. The former receives its stimuli from non-neuron sources whereas the synapses of the latter contact other, non-neuronal cells. Furthermore, there is a variety in the structure for each type of neuron in its parts. An appreciation of the possibilities and limits inherent in the nervous system's complex structure, interrelations, and functions must underlie any discussion of its activity: mind. (See Eagleman, 2020 for a broad review).

3.12 The Transformation of Information by Analog and Digital Means

Despite all the complexity in its cellular structure, there is a simple bottom line of action for each neuron; it either fires (propagates an impulse down its axon) or not. This is the digital process. In essence all the information transferred is coded, at the neuron level, into a binary form. However, in an analog process, the body of the neuron and its dendrites receive signals from many neurons (often many hundreds). At the distal end of the axon, a propagated impulse releases various amounts of chemicals, (called neurotransmitters) into the synapses where it is in contact with many other nerve cells. The complexity is further increased by the fact that incoming signals can either increase or decrease the possibility of the receiving neuron to fire. In processing information, the nervous system uses both analog and digital mechanisms. The former type of information is continuous in metric content, whereas the latter is in the form all or nothing – as are the one or zero in digital computers.

When they fire and send the impulse centrally, the specialized afferent neurons convert the analog energy input (light, sound, pressure, etc.) into an initial digital form. And efferent neurons affect, for example, multiple muscle cells, but each muscle cell is impacted by a single non-neuronal cell. The outcome for that receiving cell can be a variant of digital response depending on cell type.

After a pulse of transmission into the central nervous system itself, the complexity increases markedly in the form of the multitudinous interconnections and feedback systems. This arrangement of stimulation from many analog sources results in one binary signal. The further spread of stimulation and repression into complex networks underlies all of epistemology, a function of the brain-mind. The ontology of the brain-mind is unique because of the emergent properties of this complex, but highly structured, system.

Chapter IV
Defining Brain-Mind

4.1 The Ontology of the Brain-Mind.

All the properties from the atoms, to the cells, to the organized system called the nervous system, influence the possibilities and limits of what the functioning brain-mind can and cannot be. One cannot stress enough that the word "mind" in this context refers to a process, not a material entity. It is the result of energy flows in an organized system over time. Although they cannot be totally discounted, some of the important properties of matter/energy significant at one level are insignificant at the next level of structure, size or organization. The whole is not only more than the sum of the parts, but it can also lose, as a unit, some of the possibilities of isolated component parts. However, the emergent properties of a complex system often overshadow the properties of its component parts. To get a sense of what that activity, mind, entails, we will discuss the interaction of the various parts of the nervous system and their activities. To give some context to the ontology of mind we will first need to take a short look forward to the next part of this book, epistemology, in which the discussion of knowledge takes place

4.2 Knowledge- the Emergent Possibility of Mental Models

To this end I propose the following definition of knowledge. The thought processes that led to this definition will the subject of Part II – Epistemology.

"Knowledge: a conscious model of the world which allows a complex life form to formulate more accurate predictions of the consequences of actions and reactions of the environment and of the physical self."

If this seems to leave a lot out of the lay definition, it is by choice. I use the phrase "mental model" – or just "model" in the context of this book. It does, however, cover much of what is commonly included.

We will see that conscious and unconscious models are a part of a dynamic process. We will also discuss how emergent properties make these models possible and what the inherent limits and distortions of these models are. In Parts III and IV the discussion of mind and knowledge will be connected to the ideas concerning the interpretive and projective processes. These include the emotions, the concepts of goal and values, and the ultimate goal of philosophy, the understanding of what the sources of wisdom might be, what its limits are, and the uses to which it may be applied.

Because mind is based on the actions of a complex nervous system, the content of philosophy must be cognizant of, and consistent with, the full array of characteristics of the whole and its sub-systems. The potentials are vast — beyond any possibility of full description — but are neither "uncaused", nor "random", nor dissociated from the rest of the universe. Outcomes will be the result of the potentialities and the laws of probability.

To review: by "potentials" I mean all the relationships and actions allowed by the P-laws over time. Probability is a measure of the frequency distribution of all the possibilities, given the properties of the pertinent system at a particular time. The properties of a system are the actualized potentials of the system at that instant of time.

Given the above, mind is dependent primarily on the neuronal components, their combined structure as an organ, and the energy flows through its sub-systems. The components and their structure determine the possibilities and probabilities of the energy flow, and together they are the efficient cause (See § 4.7) of a particular mind. The physical structure of the human brain is relatively stable over time, compared to the energy flow coursing through it. This constantly changing kaleidoscope of energy flows makes the study of the subcomponents of mind especially difficult. The descriptions must therefore be, by necessity, general and incomplete. This is not a reason, however, to replace investigation with fantasy.

4.3 Scientific Studies Concerning the Nervous System – An Overview

The physical scientific investigations focus on the structure and neuronal connections of the nervous system, the biochemistry of the cells and its membranes, or the properties of the energy flows. At the higher end of the scale of emergence, investigations concern the interaction of the outputs of mind. These outputs are called by such names as thoughts, emotions, and actions. Roughly, these processes are studied by the fields of physiology, psychology and in the interactional aspects by sociology. When mental processes of more than one subject interact, the emergent properties are of ever greater complexity.

Neuroscientists study the physical subdivisions of the brain and its energy flows. This kind of information is necessary to have any chance of understanding how these components contribute to the inputs to, and outputs of, mind. This allows psychologists, sociologists, philosophers and others to tackle these interactions and combinations on a firmer basis. Mankind now has many areas of study that combine the interests of the basic sciences. They have names such as neuroanatomy, neurophysiology, molecular physics, genetics, psychology, sociology, economics, ecology and further subdivisions based on the intersections of the mother fields. Finally, modern philosophy can try to tie them together for the purpose of understanding what the potential purposes of our lives, as individuals, communities, and species could be. And should be, if we are to form more productive and less destructive communities.

Some of the areas of scientific study are not completely amendable to a prospective, experimental approach. Examples would be history (including the history of the physical universe) and evolution. Such areas require a different kind of scientific approach, namely, the gathering of data that is naturally available. From this one can form a hypothesis, and then gather further data, especially of the kind that might force us to change that hypothesis. Experimental science is difficult because it attempts to isolate and study subsystems. It has blossomed because humanity has created instruments that are sensitive to a great variety of energy flows and can take measurements beyond those that our senses can differentiate. Observational science is difficult in a different way. The source of data is vast and the amount of information is overwhelming. Therefore, we must impose categories on the data in order to manage the complexity. These categories can then function as first hypotheses.

The resulting categories are to a large extent arbitrary. But if based upon close examination of the levels and types of systems that share common emergent properties, they have a good chance of mirroring reality. And, when investigating reality, we must understand that there are no clear boundaries as to the interactions that can occur, except in ultra-simple cases. This leads to a limitation to the understanding of how the systems relate and interact in their many different ways. Each type of interaction is potentially the source of another field of inquiry. Practically, the studies are initially limited to the typical or common interactions. By typical I mean those interactions that are most probable and therefore most actualized. But the study of the atypical outcomes in a system often teaches us much more about the possibilities and limits of that system. In living our lives, we commonly use mental models based on the typical, but we ignore the less probable possibilities at our peril. Our study of the interactions of our reflexes, instincts, emotions, and rational thought must include the understanding that the typical is not the whole picture. Real, although unusual, occurrences are the reason that blanket statements, that is absolutes, should be avoided when discussing any complex topic of interest to philosophy or science. For improved communication the various fields of study must have and further develop a vocabulary that honors distinctions and avoids unnecessary and misleading generality.

4.4 Why "Typical" and not "Normal"

Speaking of a typical human I mean this. Typical as used in this book, does not mean average, ideal, standard, modal or even consistent. It refers to an arbitrarily wide mid-range of abilities or outcomes the commonalities of which can be studied, and against which the less common – atypical – occurrences can be compared and understood. I would suggest roughly the middle 80%, but that is of course open to discussion. The typical adult mental capabilities are chosen to be between the structural completion of the executive and cognitive prefrontal cortex that occurs after the teen years, and before any diminution of mental abilities as occurs in many older adults.

In contrast, the common use of the normal includes the concept of a standard of acceptability or wellness. It is also used pejoratively, an all-too-common outlook that has no place in science or philosophy. This term is therefore avoided in this book.

The fine tuning of philosophical terms for concepts is most often worked out by brain-minds which have the interest, time, and the support of their society and also have a broad understanding of the real world. When beings have these attributes, they are said to be philosophically active. Others with similar attributes, but working on other topics, are called scientists (both physical and social), linguists, artists, and theologians among other groupings. The difference between them is subject matter and the availability of data. Their abilities and contributions need not be exclusive to any one field — hence the name polymath for exceptional thinkers. Today the enormity of data from many, previously unknown, fields of study make such broad contributions quite rare. Polymaths are not typical human beings.

4.5 Vocabulary and Semantics in Philosophy: A Form of Information.
Compared to their use in the sciences the meanings of individual words (that is the concept for which the word is a symbol), as used by various philosophers and "philosophical schools", are much less consistent. This makes it very difficult for any non-professional to follow ideas presented in philosophical writings. I am not about to dictate what any specific word should mean (what concept it points to), but strongly urge that some forum be instituted that will bring order to the philosophical discussions. In this chapter I will make some proposals concerning the vocabulary at different levels of discourse.

To some degree the suggestions are based on my views on epistemology. At this point I will foreshadow that discussion, but leave the details to the next chapter. Knowledge, if it is to be used to affect reality, must have some coherent relation to that reality. This means that some way must be found to differentiate those parts of the imagination that lead to knowledge (See § 4.2 above) and those that are unsupported fantasy. To improve clarity, I will now give a definition of the word "meaning." I think that we all agree that in some sense words have meaning, actions have meaning and lives have meaning. What is the common thread that connects these usages? I propose the following.

4.6 The Meaning of Meaning
When we speak of meanings, we speak of the consequences of causes (actions). The meanings of our life are the many effects we have had, directly and indirectly on the world. The same is true for the meaning of non-human actions. The reason words have meaning is that they cause a change in the brain-mind of the listener or reader, and of course of the user. Words have

consequences, even when the speaker and the listener are subsystems of the same person.

Words point to a concept or to relationships between concepts. They are symbols and as such are a subcategory of everything that can be said to have symbolic meaning. Symbols include anything from the obvious parallels of sign language and body language, pictorial representations, and most non-accidental sounds created by animals, as well as arbitrary signs developed by humans for logic, mathematics, and other special uses such as editing.

To be a basis for accurate communication between sentient beings, the intended and actualized meaning of a symbol must be the same for both the source and the recipient of that symbol. When there is obvious confusion as to the meaning (the expected versus the actual consequences), the source of a word or phrase will say something like, "That's not what I meant!" That is, the idea or concept the source had in its mind was not the concept created in mind of the recipient by the symbols used. For good communication, the symbol and the concept should form an agreed upon tautology. Using abstract symbols and relationships, depicted as formulae, we note the following.

For any concept let X= (**a&b&c-d**), where X is the word and the phrase in the parentheses consists of the parts of the concept. The parts include the included and excluded properties. As elaborated in the next section, the components of the concept may themselves be words, pictorial mental images, memories, or emotions as well as other nervous system output, whether clear or hazy. It is obvious, that words used in any exchange of ideas, whose meanings are not shared, lead to manifold, unclear, and often mistaken understandings. In a quest for understanding, it is therefore imperative that when X = (**a & b & c -d**) it is not perceived as X= (**a & b & c -f**). The phrase (**a & b & c -f**) must have its own symbol to avoid misunderstanding. Since the two concepts have much in common, they might seem, and even be, part of a family of concepts. This leads to the second point. Short phrases or compound words will be necessary. This suggestion is based on the use of the format, expounded by Linnaeus, of genus and species. Modern humans are H. sapiens, and we are related to Homo neanderthalensis, Neanderthal man. We are related, but not the same. We share ancestors and to a small degree seem to have even intermingled. Differences are however great enough so that we need a different verbal symbol for each group.

This method, with a twist, is commonly used in everyday life and in some philosophical writings. It is not, however, used sufficiently in the latter so that clarity outweighs confusion. I am proposing, as a common method, the use of a noun modified by an adjective. The noun is the genus (close family) and the adjective is the species (individual). In English they are reversed in order from the Linnaean system, but not in Spanish — for example, "red ball" versus the "pelota rosa", where "ball/pelota" are part of the same the family. Other parts of speech such as adverbs and other nouns can be used to indicate species, but the concept of a naming system being a short phrase is retained. Furthermore, in some cases the family name and the individual indicator can be concatenated — for example the words volleyball and baseball (a method frequently and notoriously used in German).

An early example of this method is found in defining the members of the family of causes, an important aspect of any ontology, as distinguished by Aristotle in his delineation of "cause" (See § 4.7 below). His categorizations of the original individual types no longer seem adequate, but the process of differentiation by using word phrases serves a good purpose. Since cause is a central part of the meaning of "meaning," I will discuss causation in that section. A philosophical word that has been given adjectival differentiation is monism, with seven listed under the term in Wikipedia (2021). This is a good start.

There are further linguistic stumbling blocks. Part of the problem for communication is that many of the words used in philosophy go back to the ancient Greek and Latin, while others originated in various other languages. In both cases the original meaning of a neologism is often difficult to ascertain. Any language, that is, a set of symbols and their use, is difficult to translate into a new set — especially when the original verbal culture has morphed, by many iterations, over many centuries, into a new one. The problem is exacerbated when two languages do not even share the same proto-linguistic base. Also, available information, in the scientific sense, was much more limited to the ancients. However, it is also clear that their native understanding of human psychology is only now being slowly surpassed.

Translation will often affect, limit or distort the content of any concept as compared to its original meaning for which a particular word was used or coined. An example is the use of the English word "dialectic", and its earlier sources with "several specialized and largely unrelated meanings particularly when used by Plato, Kant, Hegel, Marx. (Mead, 1946, Pg. 379). The work of the first generations of philosophers must be understood with these caveats, and their conclusions should only be criticized for the use of poor logic. If this is done, their results are surprisingly good generalizations. The details of the concepts have improved, as information has accumulated, and will continue to be improved. As more information becomes available, on any particular topic, especially the kind garnered by the methods of science, new minds will need to create new combinations and permutations of the old concepts leading to the new and it is to be hoped improved ones. The definition of symbols for the changed concepts must change apace, or be replaced or supplemented by new words or phrases, through a concerted effort and agreement of the discussants in the field.

Anyone who has tried to translate from one modern language to another is well aware that basic concepts can often be translated only in approximate and subtly different but important ways. This is especially problematic in translating poetry, metaphors and idioms. Philosophical writings, which are also creative works, are also plagued not just by the typical problems of translations, but also by the use of many, often idiosyncratic, neologisms for which there is no substitute, even in the original language. The answer to this difficulty is not an insistence on who, or which camp, is "right", but rather the development of technical terms that have a common currency. At the very least, clear, adequate definitions, based on such a common currency, must be given. Only then can progress been made. Hence the many definitions and description of terms in this work.

4.7 Causation: A Physical Concept

The concept of causation is intertwined with the triad of potentiality, probability, and actuality. A description of causation tries to explain why things change as they do. Today, in a world view consistent with the methods and findings of science and logic, we understand that all change is rooted in the physics of matter/energy over time. The mathematical models used as a shorthand for the processes that occur do not concern us here, only the fact that change occurs.

There often is found the emergence of new possibilities when there is a change in the contents or relationships of a system. If the contents and relationships are composed of very few variables, with no extraneous inputs, then the potentials become the actual, over time, with a probability of one. But reality is not simple. This means that any given set of content and relationships has multiple potentials, each with a different probability of actualization. This makes the untangling of real causes and real effects from the spurious very difficult. However, sometimes the probabilities are heavily weighted toward one actualization. Then we allow ourselves to talk about the proposition that "X" is the cause of "Y". This statement of course often ignores the fact that "X" itself has a cause or causes in an infinite regress over time.

Aristotle was well aware that causation involves a series of changes, some prior to the others and some of greater significance.
Aristotle speaks of:

Material cause — loosely meaning the matter/energy involved.

Formal cause — the arrangement, shape, and structure of the material cause; that is, the relationship of the material in space and time.

Efficient cause — the agent. A carpenter is the efficient cause of the table **s/he** builds.[3] In physical terms this indicates the source and properties of the energy that acts on the material and formal causes, in this case the agent being a human being.

Final cause — the goal toward which the cause is aimed. This obviously assumed that a sentient being, with a concept of a different or desired future, underlies all change.

Determining how the system of understanding based on emergent properties and the Aristotelian ones relate to one another, and which one describes reality more closely, is not a straightforward process but a possible one. The material cause seems to relate easily to the modern sense of matter/energy as the basis of all the potentialities of being. However, the Aristotelian versions of material causation, do not seem to include a clear idea of energy as a component and variable. When there are no clear concepts of the P-laws concerning the energies that are involved in the changes of the material world, then magical thinking, (a form of confabulation) may fill the gap. Or the incompleteness is just ignored.

Einstein's famous formulation of $E=MC^2$ shows how matter (M), energy (E) and time are related. Time and space come into the equation because speed (C), the speed of light in a vacuum is the ratio of distance divided

---

[3] s/he ← Gender neutral pronoun used in this book. Pronounced 'sheh-hee".

by time. In modern terms the material causes would be matter/energy, space, and time as well as their relationships.

The formal cause states that the form – the relationship — between subunits of an interaction is a different parameter than the material cause in defining the possibilities and limits of change. Inherent in this statement is that structure is a major determinant of the potentialities of an interaction. In modern terms we see that the structure is what determines the direction, intensity, and types of energy flows allowable. Structure is perhaps the variable with the most, near-infinite, impact on the kinds of potentialities that can emerge.

The efficient cause can now be interpreted as a source and carrier of the matter/energy that leads to change. This formulation gives it some of the characteristics of a catalyst. A catalyst is a substance that changes during its involvement in a process, but reverts to its initial state at the end.

Symbolically a catalytic reaction can be written as A+B+© →ABC → D+©, where (A) and (B) symbolize initial components and (©) is the catalyst. The intermediate step symbolizes a temporary configuration involving all three components, including the deformed catalyst (C) and the final step shows the resulting new product, (D) with a return of the catalyst to its original state (©). The (D) may signify one or more products. Energy (not shown, but often symbolized in chemical equations as a curved arrow above the linear arrows) is always a part of the conversion process.

Most people will recognize the use of the word "catalyst," as in the catalytic converter of a car's exhaust system. In this case it is a form of the metal platinum that allows a more complete and more efficient conversion of unburned fuel components. In biology, the most common form of a catalyst are the enzymes. It is these complex protein structures that allow the highly efficient conversions of (A) and (B) to (D). As important as the increase in the speed of a reaction is, the changes in the probabilities of outcomes of the possible interactions between (A) and (B) are more so. Returning to the carpenter, s/he has not changed as a builder of tables after building one, except perhaps in skill and pride. His/her presence is certainly necessary to change some pieces of wood that had the potential of being a table, into a table. This efficient cause also results in a much higher probability of actualizing that table in a specific form. It is important to note that a fallen log can serve the function of the table without the intervention of a carpenter. S/he is the inspiration for the imagined clockmaker who serves the same function in a transcendent version of life and its evolution.

Such a transcendent explanation of the efficient cause of life, collapses hundreds of billions of steps, involving even a greater number of components and catalysts, into one. A transcendent being is an imagined metaphor that collapses the time component of evolution as well and is the source of all unexplained outcomes in such a system. This leads us to the final cause of Aristotle.

The final cause assumes a vision of the future. All the other causes serve only as a means towards an end. A scientific view of the world allows for a goal, but only after the actuality of a model forming mind exists. Such a mind is a necessary and sufficient source of a goal. A goal cannot be a property of a pre-mind world. Effects and consequences in that mindless world are rather an actualization of emergent properties and changing probabilities as influenced by catalysis. The real enigma is the instant of time when the so called "Big Bang" changed what was into a process leading to what is. The "Big Bang" is a major reason why we all must be hypognostic, partially unknowing beings, when it comes to an understanding of beginnings. The details of this evolution of the components of the universe are still incomplete, and contested with other versions that are no more complete.

At this point we are brought up against the difficulty of discussing ontology without an epistemology, that explains our understanding of origins. Because of the necessary intertwining of ontology and epistemology, a preliminary statement about knowledge must now be made concerning the limits of understanding.

4.8 Unknowability – the Root of a Hypognostic Outlook

A subset of the definition of knowledge, as defined above (§ 4.2) is now developed:

"Knowledge is a model of the world as a conscious thought "

Knowledge is a subtype of information. The latter can reside in many forms, from radio waves through architectural drawings, from written descriptions to concrete physical models. But, as the well-known phrase states: "The map is not the territory". Information and knowledge are real and represent a simplified and yet complex model of the thing-in-itself. This "Ding an sich" and Noumenon of Immanuel Kant, is the basis of all models that purport to represent reality. However, models themselves are also Noumena and this is where some confusion may arise. All things interact solely by the transfer of energy, and information is based on this transfer. The consequences of this fact are probed at length in the epistemology part of the book, but for now it is enough to appreciate that knowledge is a model. As a model it has its own reality and being, but it is not the original. The original is unknowable in itself as Kant stated. Hence, hypognosis- the idea that models are always less detailed than the reality they represent.

We can ask the opposite question, "What cannot, intrinsically, be represented as information?" The first limit, of course, is anything that is not part of this universe, namely, that which is called transcendent. The second limit, more of a caveat, is that only the transfer of energy becomes information. The thing-in-itself, inasmuch as it can interact with some other thing, represents a starting point. It is not information when in isolation. For purposes of science and philosophy and all other human endeavors, information is transformed into data. Data, itself a physical thing, can be conceived of as a symbol of the energy transfer. Data is also a source of further information, but it is given a different name because it, as a source of information, is only a transfer point, not the original thing-in-itself in which we are usually interested. Information can then be transformed into a model. I propose that we call knowledge those models that are available to the conscious mind. The models that are not currently in a conscious mind, but data in the nervous system that give rise to knowledge, I will subsume under the term memory. Memory itself can be subdivided, as will be explained shortly, and is only a family name of a type of brain-mind output. Its unqualified use can hide important differences between the individual members that share the family name. Other uses of the word, such as in computer "memory," minimize the enormous human-machine gap and are metaphors at best.

4.9 Neuronal Sub-Systems

We now continue our discussion of the nervous system, the collections of neurons that compose the subsystems underlying mind. Nerves facilitate the speed with which information is passed on. As noted above three basic types seem to exist.

For discussion the neural system can be divided into many overlapping subsystems. These include: coordinating circuits; simple motor reflexes; instincts and emotions (types of complex reflexes, herein named the "I-E" (instinctual-emotional brain-mind); learned reflexes (habits); consciousness (awareness); external (to brain-mind) stimulus processing by receptors leading to the rest of the afferent system; (the efferent system (muscle and hormonal management); and the most complex at the apex, thought, the C-L system. For an extended discussion of Instinctual-Emotive (I-E) and Conscious-Logical (C-L) (see § 12.1. Each level of complexity and each system is directly or indirectly interrelated. A slightly different arrangement that can be useful is the differentiation between the PAMs (preconscious adaptive mechanisms), which I define as the instinctual brain minus emotions but including habits (See § 10.9.) This is based on their ability to function as an automatic process without the recursive aspects that emotions provide to our conscious evaluations. The complex interrelatedness of the nervous sub-systems is immense, bringing about many new emergent properties. This makes their study very difficult. The investigations of, and the subsequent understanding of, mind is therefore a work in progress. But we have better data and better understanding today than yesterday, and tomorrow will bring new insights. This potential progress is covered in more detail in Part III.

4.10 Mind -- a Process
Actions of the nervous system are subsumed under the family symbol, mind. This family may be speciated into unconscious, subconscious and conscious minds, although their use has not been standardized. I suggest that a break between unconscious and the others be made between the "lower" systems of reflex and instinct and the next level of complex reflexes, emotion. This is somewhat arbitrary but is based on the following considerations. The first consideration is based on which organisms in the phylogenetic tree have what level of nervous system. Jellyfish coordinate their movement by a simple system that keeps the undulations of the bell in synchrony. They also have reflexes that are the reactions to external stimuli. Their simple nervous system is ring-like and does not have a central core of neuronal concentration (called a ganglion, plural ganglia). A ganglion is the simplest form of a nerve cell cluster, with the next level of complexity being called a plexus. Humans have ganglia and plexuses, all of which are involved with reflexes and the processes of the autonomic nervous system. The latter modifies and controls the basic functions of the body; blood pressure, heart rate, gut motility, and so forth. One such structure is the solar plexus (celiac plexus). It is located behind the stomach, is made up of and connected with other ganglia and helps to coordinate the unconscious functioning of many internal organs. Its function

is not controlled by the conscious central nervous system although it may be influenced by it. More complex inborn reflexive behaviors have been given the term, instinct.

4.11 Nervous systems (NSs) that function as a conscious brain- mind

I wish to reserve the term conscious mind to a level of nervous system action that has at least two functionalities. The first would be consciousness, a term that will be more closely discussed in under epistemology, § 5.15. The second is the capability to non-reflexively control responses to the environmental stimuli based on internally created mental models. A system capable of only pure reflexes will be referred to as a preconscious nervous system. The reflexes responsible for instincts and emotions are therefore technically part of a preconscious mind. But, when these are part of a larger NS that includes consciousness, with which it is highly integrated, the systems may function as one. However, to understand this more complex system, we will find it useful to first discuss them separately.

Before the neurons evolved, cellular organisms had other mechanisms that allow them to react to their environment. All are based on physics and chemistry. The best known are the tropisms. Plants do not "know" what or where the sun is, where the center of the earth is, or that they need water. It is the physical response to sunlight, gravity, and the relative concentration of water and nutrients in the environment that control the reaction of a plant. Single celled organisms have a form of tropism called taxis. Directional movement not caused by the external factors is called kinesis. The advent of neurons allowed these functions to be monitored and activated from a distance.

Simple motor reflexes such as the "knee-jerk reflex" coordinate primitive responses of the neuro-muscular system. Specialized rhythmic motion such as the heart's pumping, and the coordinated undulations of the gut for food propagation is organized by locally active neuronal tissues. Both of these can be influenced by "higher" —more complex and later in evolutionary terms – nervous system circuits. These synchronizing local circuits and what they control are usually not termed reflexive because no immediate external (external to the organism) stimuli are normally involved. The gut is a special case because the lumen is technically outside the human body. (If you swallowed a long enough string, you could have one end protruding from the mouth and the other from the distal orifice of the digestive tract.)

When we speak of instinct and emotion, we refer to the integrated automatic responses to stimuli received by the classical senses: vision, hearing, touch (including proprioception), smell, and taste. A brain-mind may also create virtual models from internal stimuli (such as memories and dreams), and similar responses may ensue. Together, pure instinct and pure emotion, are mechanisms that can therefore be thought of as complex reflexes that involve more than a simple stimuli-response loop. The instincts are the next step above simple motor reflexes, and they involve the complex coordination of motor responses. The need to separate out instincts and emotions versus cognitive evaluation will be addressed in Part III, § 8.6, with the discussion of purpose.

Emotions are more complex, but nevertheless are still preconscious in origin and are automatic. They may include motor responses, but add to this the activation of the body's hormonal systems and chemical responses in the amygdala (an anatomic area of the brain). What does it mean then to say that we control our emotions, if they are nothing but complex, but nevertheless still preconscious activities of the brain? There are two aspects to such "control," though a better word might be "influence." In the first case, the influence is post hoc. Once we are conscious of a personal emotional response, we can consciously dampen this emotion provided we judge the response to be harmful to ourselves or others. Secondly, although we cannot directly turn on any emotion, we can use internal stimuli. We can consciously imagine circumstances, based on our memory, that have in the past created the desired emotional response. We also can imagine new causes, using "second thoughts", concerning a current situation if the original emotional response seems erroneous to our conscious mind. In sum, a cerebrally initiated, conscious mental scenario can act as a stimulus likely to induce or modify a desired emotional response. This will be important in topics in Part IV such as "free will."

An example using the above-mentioned "knee-jerk" motor reflex will help to clarify the difference between the conscious dampening of the emotions and their alteration by internal stimuli. First, we can control, consciously, those muscles that move the leg at the knee joint to prevent this reflex. By this method we can voluntarily block the ability of the reflex to occur. Second, we can cross our legs in a sitting position and relax the muscles consciously. We can now, again consciously, take a tool such as a small hammer, and tap the patellar tendon just below the knee-cap. Now the leg extends (kicks out) involuntarily, in a facilitated reflex. The reflex itself is not voluntary, nor is it the same as voluntarily kicking with the leg.

We now can see how we have induced an involuntary motor reflex by using conscious control over a different set of muscles and nerves. We can call this a self-induced motor reflex, indicating not control of the reflex directly, but an indirect provocation of it.

The parallel with emotions should now be clearer. Reflexive emotion refers to a response to external stimuli, unmodified by the conscious mind. Self-induced emotional reflexes are activated by using other circuits than those afferently stimulated. Such circuits can include the voluntary control of, for example, our facial muscles to produce the complex motor activity known as a smile. This voluntary smile can induce a positive emotional response which is similar to the automatic emotional response that may then, secondarily, create a similar smile. Both are based on a feed-back loop. The temporal order is, however, important in differentiating those two situations. We also can call forth a memory of a time when an incident brought forth that reflexive positive feeling and a smile. This can again create the emotional response that creates the muscular activity of an involuntary, although perhaps desired, smile.

Parallels can be drawn across the emotional spectrum, but do not need to be itemized here to make the point that there are purely reflexive emotions and those self-induced, but nevertheless still reflexive. Cognition can be a tool that creates a situation of initiating a desired emotional response. But then it is not a sensory induced (pure) emotional reaction.

Even though we are not always aware of them, I exclude moods, emotions and instincts from the neuronal activity that is the unconscious mind in humans, because these automatic functions are usually so closely integrated with the higher cognitive functions. These activities are major aspects of the preconscious mind, that is easily and often automatically integrated with the conscious mind. What we must not forget is that, when emotions occur, they can be initially a pure and primary response, but that this is rarely the case because of the close interaction with the conscious mind. This understanding becomes critical in Part III where we will discuss "evaluative mind." A complex system that gives rise to such concepts as beauty, morals, ethics, justice, good and evil.

4.12 Memory -Laying Sown and Retrieval

Another major aspect of the brain's function is memory. I exclude the physical basis of memory from my definition of mind (it is part of the physical brain), but include its laying down and later retrieval as functions of mind. These nuances will be outlined at length in the following chapters, but I foreshadow that discussion here so that those chapters may be read within an organized context. For this purpose, I offer a minimal definition of the family concept, mind.

"Mind is the functioning of a nervous system. It makes use of the physical substrate of the brain. It is not a physical thing, but is an organized activity."

To clarify the distinction between a physical system and an action, I offer a simple metaphor: a horse or motor race. The horses or cars, the racetracks or ovals, the movement (matter in a space/time relationship) of the racers or the fact that it is a competition (another activity) — none of them in themselves constitute a race. The race is the activity that combines these parameters in a specific way but is not, in itself, a physical thing that has metric qualities. It is the same with mind. The neurons stand in for the horses or cars; the underlying supportive structures and relative relationships of the neurons are the spatial equivalent of a racetrack; the chemical and physical changes that lead to the firing of a neuron, and the firing itself, can be seen as parallels to the motions of the racer; and the main use of mind is to manage the needs of life, while the activity of a race is based on the concept of competition. When a race is proposed, it can be said that the brain-mind is purposefully used. But mind itself, as an activity, is not the result of such a proposition. It is rather the emergent result of the evolutionary process that enhanced the continued survival of a specific living thing and, indirectly, its species. Mind, like reproduction and metabolism, need not be the result of an a priori, consciously arrived at, plan.

The whole purpose of this chapter has been to show: how evolutionarily old, functionally basic and structurally diverse the human system of nerves is; that it has become more and more complex; and that the emergent properties arising out of these complexities lead to the human condition — the main subject of philosophy. This chapter has set the stage for the subsequent parts of the book that are more central to philosophy. It has outlined the fact that all complex systems are based on the combinations, interactions, and increased numbers of simpler systems and that the P-laws governing the properties of those simpler systems are still in effect. Indeed, this pyramid of change leading to mind is only the result of innumerable cause-effect relationships at different emergent levels of complexity. The levels give rise to new possibilities and probabilities and thereby give rise to actualities. Such actualities as the emotions of love and fear are produced by the nervous system. So are the relational concepts (models), such as justice and free will. But all have their roots in the P-laws and the emergent properties studied by the classical "hard" sciences. How we come to "know," classify, use, and evaluate and how we are influenced by these mental actions in the course of our daily lives is the subject of the rest of the book.

4.13 Divisions of the Brain and its Effects on Philosophy

Before we leave this chapter, a few words about the physical brain will round out our understanding. The human brain has three major divisions, each of which allowed for a new level of emergent possibilities in interacting with the environment. All of them are active simultaneously, although with differing levels of dominance in any particular case. A version of this description was the triune brain promulgated by Paul D. MacLean (1990) and earlier by Carl Sagan (1977). Roughly speaking this divided the brain into the brain stem (reptilian complex), the mid-brain (paleo-mammalian complex) and the cortex (neo-mammalian complex). The later can be further functionally divided into primary, secondary and tertiary areas – each one with new levels of modeling capabilities. All of these anatomical divisions are now known to be an oversimplified, but important approximation, of the functional units of the central nervous system. The evolutionary structural developments and the much more complex interactions of these anatomical areas are the basis of the emergent properties discussed in the third part of this book, "Evaluations". The import of their activities will be developed in discussions of such subjects as beauty, free will, justice, evil and purpose. These evaluative activities of mind cannot be discussed adequately without some basic understanding of the underlying structures and their activities as they relate to each of these concepts. The discussion of these levels of emergence, although fitting into this section of ontology in terms of new abilities, cannot be described well without delving into the actual functions they support. The topic of ontology of mind is therefore continued in Part III. This organizational conundrum has been one of the more difficult aspects of writing about what the yearning and search for wisdom is all about. We will therefore continue by discussing what "knowing" seems to be about, under the classical heading of epistemology.

PART II -EPISTEMOLOGY
The Study of Knowledge

Chapter V
Knowledge: A Mental Model

5.1 The Connection with Ontology
In many ways epistemology is central to the ontology of a conscious mind. Knowledge presupposes that there is something to know. This includes the knower — such as René Descartes, who is famous for his "I think, therefore I am". The process of knowing, as described in this section, is consilient with the infrastructure of matter/energy and space/time. Epistemology is an examination of a structure-activity complex, the emergent brain-mind, a process which is recognized in a few other, mainly mammalian species. To advance these studies, the ontological aspects will need to be part of any investigations of epistemology, interwoven with the mental mechanisms of "knowing". To know is also the connection between the ontological bases and the attempts at wisdom as discussed at the end of this section.

5.2 What Does it Mean to" Know"
The verb "to know" is one of the most polysemous words in the English language. Polysemy, the multiple referents of a verbal symbol, and the associated problems in philosophy is well discussed by Bunge in his essays (Mahner,2001). "To know" has many, more or less related, variations in its possible referents. Misunderstandings and confusion in discussing this topic reign because there are so many common-language and technical meanings.

There are two major reasons for this. The first is that the speaker and the hearer (or writer and reader) may have different concepts in mind when they use the word "know". The same is true for the word knowledge. The second major kind of misuse occurs when it is used anthropomorphically, implying conditions and causations not inherent in the ontology of the system under discussion. I have come to the conclusion, as will be explained below, that the basic, root verb "to know", should be reserved, especially in philosophical and scientific discussions, to the following concept definition.

To know is to form and then hold in consciousness a model of parts of the world. It allows a complex life form to formulate more accurate predictions of the probable results of interactions in the real world. This involves the self and the environment in all its permutations.

Unfortunately, its use in every day discourse will continue to have vague referents.

To be useful, "to know" also must mean that the currently held mental models are believed to be the most accurate and consistent representations of reality available. If the perceived likelihood of a specific situational outcome is great, as attributed by a knower, the stronger will be their belief in the correctness and completeness of their mental model – their "knowledge". This ability to form and accumulate mental models, when based on accurate data and sound logic, has been critical and necessary for the success of the human species in its struggle to adapt and survive.

5.3 Knowing and Surviving

H. sapiens and its immediate ancestors were and are relatively frail creatures, whose existence, without the ability to know, would be limited to a relatively benign environment. This environment would have to have: easy access to food for energy and maintenance; easy access to clean (non-toxic and noninfectious) water; a narrow range of climate; and few if any environmental dangers, whether predatory or other sources of severe trauma. All living things are limited in some way, and therefore owe their survival as a species to their ability to adapt to specific environments. When it is not possible for a particular species to thrive in new or changing environments, other species or genera may overcome this difficulty and replace the former. Speciation, that is subdivision based on genetic variation, allows for the spread of the common genes while increasing the range of livable environments. The more successful species will then pass on their special and their common genes. These parents of future generations, initially slightly atypical in their genetic structure, were then able to continue life and reproduction in new or changed environments. It must be recalled that these genetic changes are a priory, in both a temporal sense and in that they are not teleologically driven. To speak of genetic adaptation in nature is therefore inaccurate. Also, the word selection is misleading. What occurs in speciation is an improved fit of a particular, already present, genetic pattern to a particular environmental condition, thereby improving the chances of survival and reproduction. Both adaptation and selection imply that the choices were made, by beings with the use of mental models of a future. Knowledge is the tool that makes choice possible and has positive species survival value.

Some genera, rats and cockroaches come to mind, have broad abilities that are a workable fit in many situations. These abilities include an omnivorous diet, resistance to infection, instincts to avoid toxins by smell or taste, and an inconspicuousness or an ability to hide. There are many other characteristics which allow a broad range of environments to be habitable, but humanity has only excelled in the first ability from this list, the rest are fair to poor. Infectious disease became a bigger problem for humanity as more contact between different tribal subgroups became more common. The horizontal spread of pathogenic organisms from person to person has historically been a major cause of decimation for mankind. Both our senses of smell and taste are minimal by mammalian standards and our bulk makes us quite conspicuous. On the other hand, our excellent vision is a plus. A potential negative is the long period of time between infancy to adulthood, creating a greater need for energy to just reach a reproductive age. All the negatives were, to a great degree, overcome by our instinct to be a social species and our ability to know.

Knowledge made a difference! With the greater ability to anticipate difficulties and dangers, and preparing for or avoiding them, not by instinct, but by more flexible and exacting model making, we gained a novel edge in the competition for survival. One result was the construction and use of above ground shelter for protection against the elements (simians are not good at burrowing). When it was not possible for a particular group to continue to thrive in new or changing harmful environments, such innovations as clothing, housing, and defensive perimeters changed the odds. Migration was driven more by imagining a better place than by pure instinct. With knowledge we now can now speak of purposeful adaptation and selection. Choosing a viable location to live in as a group, could be based on previous successes and failures, which depended on the models that included an access to long-term memory, both personal and that held by at least one other member of the group. The often-mentioned adaptation of fire for uses such as cooking, illumination, and finally to metallurgy is another important example of the usefulness of knowledge. The learned and passed on knowledge of herbal medicine contributed to our fight against injury and disease. Of course, we also continued to rely on and use our basic reflexes and instincts. However, conundrums arose when knowledge and the more primitive neurological reactions to the environment clashed.

One instinct critical to the development of knowledge is that of inquisitiveness, that is unplanned searching of the environment — a trait shared with all mammals, and to a lesser degree with other animal phyla. Instinct directed movement improved on randomness, tropisms, and chemotaxis which is used by more primitive, less complex animals. The latter are all reactions to stimulation by various forms of energy or to chemical reactions both internal and external to the cell. No nerves are involved. Unguided (by knowledge) searching behavior can also lead to the dangers implied by the saying "Don't stick your nose where it doesn't belong". On the positive side of the ledger, habits and instincts based on fear can help avoid future dangerous explorations and searches. Conversely, based on previous personal experience or communication from others, the chance of finding something useful for oneself or rest of the group in an unknown environment is increased. The various types of interactions between levels of existents (things that exist) are found in Table I, § 7.7.

In short, H. sapiens must know, as defined, to survive. It is this kind of knowing that philosophy is concerned with. For the purpose of fruitful philosophical discussion, clarification of this concept family's subcategories must be created, by using more precise terms for the various concepts that this common word refers to. Other languages have their own polysemous words and problems with concept identification. This can make the translation of terms very difficult. Sallis (2002) discusses translation in very broad terms. He gives many examples from Greek to English, as well as many examples of the translation of Shakespeare, especially to German. He provides extensive examples of the difficulties of translating meaning versus vocabulary. This is why scientists use scientific terms that are the same across languages.

Philosophy needs to do the same. This is the work of professional groups or autodidacts. But first, an individual must make a suggestion.

5.4 The Lay Uses of the Word "Know"

Some of the more common lay concepts invoked by the word, know, need their own terms in philosophical and scientific discussions. The greatest misuse is when the verb is used in an anthropomorphic way for either non-living objects or living beings that have no nervous systems. Then there are those uses that refer to unconscious mechanisms such as instincts and reflexes. Closer to the narrower meaning is when it is used for cases where only parts of the nervous systems are used. The most common missing function is the retrieval of experience-based memory that brings previously formed concept into consciousness. Finally, we have the use of "know" where the broader term "believe", which includes more than what is known, is used.

This outline is hardly self-explanatory, but is needed for context in reading the rest of this chapter. What is involved when we speak of knowing in a pure epistemological sense, and the other lay terms, will be expanded on in the next chapter, especially § 6.6, with examples and further discussion.

5.5 The Relationship of Epistemology to Ontology

Whereas general ontology has been greatly informed by scientific studies, the "black box" of the brain-mind has been, until recently, resistant to direct study. This topic has been amendable, historically, to investigation by two main methods. Observation and manipulation of the behavior of sentient animals (including us) and introspection. Although introspection is a completely private matter (unobservable by others), the results can be communicated, compared and discussed. For millennia, behavioral observations, both formal and informal, have been used as a basis to manipulate future behavior, but the underlying nervous system processes have been totally opaque. Advances in neuroscience have changed all that, and this progress is and will be essential to discussions of what we mean by words such as know, think, feel, imagine, compare, memory and recall. As noted before, the fine details of the scientifically garnered data are not critical to a philosophical understanding, but the structural and relational rules of a scientific brain-mind ontology must be considered and incorporated.

5.6 Structure of the Nervous System, Continued

A more detailed discussion of the physical nervous system, as a system of specialized cells is now in order. This is to supplement the discussion already initiated in the chapters on ontology (esp. Section 4.3). Although all neurons have dendrites, cell bodies, axons and synapses they have specializations dependent on their functional location in the brain. Some influence mostly other central neurons whereas others support bodily functions.

Afferent neurons are those that react to stimuli from non-neuronal sources. They are the carrier of all stimulation from the external environment and the rest of the body. Their activation is the only source of information upon which knowledge is built. Classically this includes the five major senses of vision, hearing, touch (pressure), smell and taste. To these can be added those neurons that respond to temperature and to damage. Of note is that all kinds of input can lead to the sensation of pain or discomfort if the level of stimulation is high enough.

The best studied of these senses, in humans, is vision. It is the sense most involved in judging spatial relationships and differentiating between the wavelengths of light in the visual spectrum (color). The sense of color, is created by information garnered by specialized cells of the eye, called "cones", concentrated at the center of the visual field. Another type of cell, called "rods" are more peripheral and are the source information that is gathered by reacting to a lower intensity of light.

The activity of neurons sensitive to sound, usually pressure variations in air, are useful in several ways. The best-known use is communication at a distance: whistles, calls, grunts, and of course speech. Sound is also used for determining spatial relationships in bats, although rudimentary use in this way has been described in blind humans as well (See "Human echolocation" in Wikipedia). Another less recognized use of sound is in the judgment, and evaluation of time, especially that of short duration. The most obvious use is rhythm in music. Before the advent of any mechanical clocks such rhythmic tapping was all we had to judge the passing of short periods of time. Longer periods of time seem to be functionally evaluated by the order of occurrences as recorded in memory. Light and sound together, when incoming from the same source, can be used to estimate relatively large distances. The disparity in time between the perception of lighting and the ensuing thunder is an example.

Gravity is sensed by pressure and is also involved with spatial orientation and the sensation of movement. These senses and others, interacting in complex ways, are the basis for classical psychological studies.

Efferent neurons are those cells that send information to non-neuronal cells, mostly those in muscle and hormonal systems. They are the last neural link in producing behavior, and are necessary if we are to have any effect on the world. Without efferent neurons our bodily functions would not be properly coordinated, and we could not communicate with others at all.

The majority of neurons do not fall into either of these categories and have been called central neurons (or interneurons) since both the dendrites and the post axonal synapses of each neuron are contained within the central nervous system (CNS). They are the majority of the neurons whose bodies reside in the skull enclosed brain. These neurons can be both the senders and receivers of stimuli to each other. These "internal stimuli", which together with those arising outside the NS, the "external stimuli" comprise "all stimuli". There are other neurons that act in the production of reflex arcs, called relay neurons. They act as intermediaries between the afferent and efferent neurons outside the CNS. They belong to these relatively simple systems and are not directly involved in the higher functions discussed below. Together, these various neuronal cells and their supporting non-neuronal cells, such as glial cells and those of the circulatory system, can be seen as the hardware which is the material basis of mind and all its functions.

The CNS is where the most complex action takes place and it is a very difficult system to study. This great complexity and the lack of direct contact with the non-neuronal world has given rise to the sense of mystery surrounding the human psyche. How could one have said more in the past than that the mind is mysterious? The first approach to better understanding is by looking at brain-mind function in a general way. The second is to develop a more exact vocabulary in discussing the functions of the systems we are aware of. The latter will be discussed function by function, using underlying ontological, relational, and physiological concepts that connect them all. To achieve this, I will first discuss in broad terms what nervous systems can do. For this I will utilize two basic concepts: abstraction and imagination

5.7 Building A Mental Model. Abstraction – Definition and Example

I will begin with a definition of the lay term, abstraction, which highlights the commonalities between various types of usage. We speak of the abstract of a scientific paper, abstract art, and an abstract idea. Merriam-Webster.com defines the noun as: "a brief written statement of the main points or facts in a longer report, speech, etc." and the verb as: "relating to or involving general ideas or qualities rather than specific people, objects, or actions." Neither definition, emphasizes the basic fact that something which is abstract, is an incomplete part of the whole. That incompleteness is the essence of all abstraction. But it is not enough to define its meaning. The second central part of a definition, if the concept of abstraction is to have any usefulness, is that the abstract must have a definable relationship to that whole. Both of these aspects underlie all the members of this family of concepts and its derivatives.

Mental abstraction, the species that is referred to in this book, is the first limiting and organizing process in the overall function of our nervous system. Basically, this means that only a small, selected part of the energy flux that impinges on our afferent nerves is utilized. The visual system, to be discussed in some detail, is a good example. The main point will be, that however we conceptualize knowledge, it is certain that it is based on an incomplete data set. The constant appreciation of this basic, necessary incompleteness is critical in formulating all the concepts that involve nervous system activity. Thus, I agree with M. Bunge (2003, pp. 242-3), that all knowledge of reality has to be conditional and will always be, at best, an excellent but approximate abstract of reality. I have chosen to give this concept the name of hypognosticism. In this book the single term, abstraction, will always be used for the phrase, mental abstraction, unless otherwise specified.

5.8 Vision as An Example of Abstraction

Let us imagine before our eyes a chair, made of wood, stained and slightly worn. Do we see the chair? Not really. What we do see, namely that which stimulates us, are the energy waves of the visual spectrum. These are not a property of the chair per se – but an interaction of incident light upon it, and the chair's own qualities which govern which waves are absorbed and which are reflected. These relationships between incident light and the individual molecules and atoms of the chair are lawful, and thereby create a lawful abstracted reflection that can pass on information concerning the structure and character of the chair (as well as the light source).

Next, some of the reflected rays of light now impinge on our retina. (Here we ignore the relative minor effect that air, cornea, lens, and vitreous humor have on the light as it passes through them.) The information contained in the light reflected by the chair is, however, only the tiny portion that reaches the eyes. At the retina, light waves interact again – this time with the chemicals in the sensory structures, the rods and cones, of our eyes. Here are created new lawful abstractions by converting, in a selective way, radiant energy into chemical changes. These chemical changes then cause a further lawful cascade of reactions in the nervous system. This starts with the optic nerve's afferent neurons, originating in the retina. The energy is passed on via other specific neurons, to multiple areas of the brain, the most important afferent destination being the visual cortex located inside the back of the skull. When the processed information then is sent, by other pathways and interactions to our short-term memory, it ultimately results in us "seeing the chair". Seeing then is actually a set of lawful interactions of energy and matter, resulting from a series of cascading abstractions starting with the interaction of light with the chair. It is the totality of these structured abstractions of our brain that cause and allow us to see and, by further processing, know, understand and be conscious of this chair.

By scientifically analyzing the cascade of lawful abstracting interactions, neuroscientists approach a better and better approximation of what it means to sense the reality that is this chair. The concept of chair is further expanded by combining the directly visually abstracted information with the help of modern tools that extend this sense. Microscopes, dyes, various lab techniques and MRIs are but a small list of the tools that impact what information becomes available to the visual abstraction process.

Further ways to experience the chair is by touching and lifting it. Information is garnered via the touch receptors in our skin and other sensors that respond to resistance to movement. Together with our visual sense our brain then creates a more complete model of the chair – a process I subsume under the term imagination. This is only the beginning of the complexity of the relationship between us "knowers" and the remainder of reality in which we are lawfully embedded.

5.9 Imagination and Confabulation

The term "imagination" is used here to refer to all the activities of the CNS that are used in building all models. These models may or may not be consilient and consistent with reality. Usually, besides the information available from the abstracting process and from the contents of memory, confabulation is a normal part of this process. It is the subset of mental processes that attempts to fills the gaps in a concept where information is missing or contradictory. It is an automatic process that often achieves the creation of an internally consistent model. Medically it is spoken of in brain damaged persons – but that is too narrow a setting – if the most obvious. If done consciously, it could be seen as "lying to oneself" – but true confabulation is not immediately recognized as a source of a full picture(concept) that is partially false. At this time the details of which subsystems are involved in "normal confabulation" is not well studied.

Confabulation should not be confused with hallucination, which is a disorder of the typical processing of afferent stimuli. The resulting output is often not recognized as being discordant with a model of reality based on the usual imagination process. Many chemical substances are known to create this effect by changing the balance of normal neuronal function. Nor is confabulation the hearing of voices or seeing of visions not based on current sensory input (in contrast to receptive hallucinations). Although these phenomena can be associated with complex disorders such as schizophrenia, this need not be the case. The mind-brain can itself initiate this phenomenon by the atypical firing of neurons. It can be loosely compared to a muscle twitch that is not the result of any external stimulus or a cardiac arrhythmia caused by atypical pacemaker cells. The importance of these occurrences to philosophy is that it can muddle the discussions on what is knowledge. If the dissociations from normal cascades of energy flows through the brain-mind are seen as a reliable basis for modeling reality, then anything can be called knowledge. That is not useful for progress.

5.10 The Functional MRI (fMRI) –Tool for Studying Brain Function In Vivo

In recent years the use of fMRIs has greatly increased our understanding of normal and atypical brain function – the brain-mind. This has been especially important to neuropsychologists. Originally MRI (magnetic resonance imaging) was used for a detailed evaluation of brain structure. Functional MRI gives us a picture of which parts of the brain are most active during specific mental activities. The most common method measures the oxygen in blood flow, which is known to be coupled to differing energy needs in areas of high versus low brain activity. This technique is so productive, that new studies are coming out with regularity. The basic methods and characteristics of such scientific studies are as follows. Firstly, the subject is awake and alert and able to follow commands. Secondly, a specific protocol is followed so that specific functions of the brain-mind are (hopefully) isolated. Thirdly, there are multiple subjects, using both control and experimental conditions. Altogether, the findings of these studies point to the following conclusions.

A — The areas of the brain that are activated under specific circumstances are the same in the experimental subjects and can be differentiated from the results seen in the controls.

B —The areas of the brain that are activated correspond and reinforce the results of previous studies based on electrophysiological studies (direct measurements of electrical activity, often during neurosurgical procedures).

C —Previously unsuspected circuits have been shown to be involved. These additional circuits indicate a greater complexity of interaction than was previously imagined.

D —Activation of similar areas, under different experimental conditions, points to the multiplicity of functions a particular circuit may participate in.

The above conclusions, with other pertinent data, indicate that the models created by the mind-brain are generated by using the abstracted information from the environment which is then passed on to and processed by highly structured, highly interconnected circuits that form the models. This complex mechanism is how a mind-brain conceptualizes reality.

The complexity of this process cannot be overstated and is impossible to grasp in full, but a general overview of the kinds of systems and relationships can be made. In formulating philosophical concepts concerning knowledge, the connection of reality to such complex, organized systems of interactions and interrelationships must be kept in mind.

A common concern is that the relationship is at best a correlation, not an indicator of the mind causing the models. More data is needed concerning, especially, temporal issues to clarify which stance is more correct. But, correlation of the physical activity of the brain with the self-reported awareness of that activity is a strong start toward understanding which parts of the brain are commonly active in similar circumstances.

5.11 Levels of Brain-Mind Complexity

The "simplest" systems are mostly influenced by external, non-neuronal, afferent stimulation. These systems, directly create a "map" abstracted from the pre-structured energy coming from the source. In this context pre-structured means organized energy flow from the external source to the sensory receptors. If the energy is supplied in a non-pre-structured or ambivalent way, our more complex brain-mind will try to, if possible, provide a structure. These 'models', of structures and relationships that don't exist, have been given names such as confabulation, delusion, illusion, hallucination, mirage, myth and others. Such fantasies can be a source of unwarranted certitude, entertainment, misunderstandings, arguments, psychiatric illness and even war. "Any relationship to reality is merely a coincidence" – as novelist like to say. Often, this caveat is not taken as seriously as it should be for the most consilient understanding of reality.

The next level of complexity can be exemplified by our sense of spatial orientation. This usually involves input from: the visual system; pressure sensors on weight bearing skin and joints; and by the pressure exerted on the cilia by small stones, the otoliths, of the semicircular canals of the inner ear. The latter although physically a part of the inner ear is not used for hearing. When stimuli supply consilient information we know "which way is up" in relation to the earth. When these senses supply inputs that cannot be coordinated by the circuits involved, we feel dizzy, nauseous and disoriented. This response of the nervous system to contradictory spatial information is a clue to the source of the feelings of disorientation, confusion, anxiety and other disharmonies when other various sources of information cannot be organized into a single consistent model. The opposite, the harmonious feelings created by consilient reality based (or fantasized) models, underlie such concepts as contentment, love, beauty and peace. All of these feelings involve emotional responses more than intellectual evaluation. The interaction of the two evaluative systems (C-L and I-E: Chapter XII and Chapter IX, Table II), which created different responses to concurrent inputs (all stimuli), creates the tensions that underlie the difficult discussions of the higher order concepts such as goodness and evil, justice and unfairness, beauty and ugliness, and all the other dichotomies that occupy our thoughts in the search for understanding that is needed for wisdom as discussed in Chapter VIII.

5.12 What Can We "know?"

Knowledge is a mental model, an instantiation, that results from the process signified by the verb 'to know':

Knowledge is the totality of the mental models of the world. It includes all secondary storage and transfer media from papyrus scrolls to the dot-com sites. Together, after retrieval from these storage sites, this information can help any complex life form to formulate, in a conscious state, more accurate predictions of the actions and reactions of that world. It is an approximation of reality, and can be conceptually wrong for both reasons of data and processing.

We now have to discuss the concepts that the critical words in the definition invoke. As noted previously, when there is concern about possible misunderstandings, definitions should be offered. It is in this spirit that the following descriptions are given.

Mental model: a unique set of activations in a neural network. The components of this activity are: a relatively stable, communicating, neuronal network creating specific patterns of individual neurons firing in that network. The resulting release of neurotransmitters causes transient effects in that network. Any particular thought is fleeting, but the constant feedback loop process can link them, forming a fuller, temporally stable, picture. Unless otherwise noted in this book the unmodified word, model, refers to the mental model.

In short, a model is the temporary activation of a particular network of neurons. When one considers that each neuron has an average of 7,000 connections (Wikipedia – under Neuron) it is clear that each model, each thought, will be unique. Nevertheless, although each model of the above-discussed chair may be unique, these models form a tight group, compared to the models of a bench. Together they form another group containing the concept "seating furniture".

This is all based on the ontology of neuronal networks. It will be recalled that the physical structure develops from a genetically controlled "architectural plan" that is realized during embryology, all of childhood and also early adulthood. The environment, which includes real chairs of different types, continues to influence the frequent reconfigurations of the physical CNS structure that underlies a particular model of reality throughout one's life.

Whether Plato's preformed ideals can exist in reality is highly disputed. A perfect mathematical structure like a circle cannot exist in reality. In maths, in a plane all points distally equidistant from the center (with a value called the radius) form the circumference. Any material circle's circumference has some width and irregularities. Is the radius to be measured to the average inner, outer or median distance? The answer is arbitrary, depending on the need for the measurement, and, the object's perfection is a value judgement based on the use of the object. Absolute, mathematical perfection is not possible. The same logic applies to perfect mathematical triangles versus real ones, and so on.

The blank slate from Aristotle to John Locke is also a misleading concept as discussed in S. Pinker's book (2002) of the same name. It is the idea that the mind is a passive receptor of stimuli. But we now realize that the structure of the brain limits the possible "writing" on its slate. However, typical but unique slates will process input in a similar way. This is why we can agree on most things concerning the real world.

To summarize: knowledge is based on models arrived at from stimuli originating in reality. The stream of causation initiated by the physical phenomena in question is processed along a typical nervous system cascade. But this does not happen in temporal isolation. Such isolation is unattainable because the model is constantly readjusted by comparing the current input to those from the memory created by similar inputs in the past. Furthermore, we share very similar nervous systems.

5.13 Belief, Faith, Certainty, Knowledge and Fact

Knowledge must be differentiated from belief, because the former is a subset of the latter. A belief is a model whose accuracy may be rationally questioned by the believer or others. Beliefs are mental models based on: scientific data (atoms), personal experiences (apples), unverifiable imagination (apparitions), and received information from others (any of the first three). It is differentiated from faith and certainty in that the suspicion of erroneous information is still acknowledged. Certainty is expressed by claiming the knowledge is warranted because there is no contradiction with the P-laws or other equally supported information. It also assumes no conscious denial thereof. Less supportable as knowledge is when a belief leads one to ignore acknowledged ignorance in the face of a need for immediate action. Insupportable are those beliefs that ignore contradictory models, especially when the emotional support for the belief suppresses the search for further knowledge. This kind of willful ignorance is all too common. The high frequency of this conscious suppression or avoidance of new information in everyday life, politics, formal justice, business propaganda, etc. led me to formulate the neologism "wignorance". This condition is thankfully less frequently involved in the formation of concepts leading to the pronouncements of scientist. Philosophers are unfortunately also not immune. Wignorance, resulting from emotional conflict and/or intellectual laziness, is one of the more pernicious reasons for a lack of progress in any field of endeavor. To continue to claim knowledge, without reassessment, in the face of new contradicting direct experience, data or a supported communication from another person does not advance the search for wisdom.

When, in common discourse, we speak of facts, the believed to be accurate models of reality, we may be speaking, unconsciously, at three levels of certitude. The first is knowledge, the second, is belief, and the third is belief presented as knowledge. Beyond this are lies. In this regard, the common use of the term, fact, is a way to claim a consilience with objective information. What the word "objective" refers to is, however, notoriously difficult to define. In its simplest form it refers to something that is measurable by physical means, and is data without interpretation. When the word, fact, is used to describe a mental model, it is closely associated with the term "conceptual truth" as discussed in Sec 6.1.

5.14 Fantasy – the Basis of All Models

Everyday fantasy is the mental creation of a model that is partially based on reality supplemented by non-pathologic imagination. It results in models or stories that are not in concordance with demonstrable data, or include explicit data that demonstrably cannot be available – like Bertrand Russell's teapot between Earth and Mars (Wikipedia, "Russell's teapot", 2021).

Fantasy has similar subdivisions as the created models based on reality, but in this case with a greater relative amount of confabulation or creative imagination. Models that are reported as fiction, partial or total, are acknowledged to be the result of (mostly) applied fantasy. However, preliminary ideas, based on acknowledged strong imagination, are very important initial steps in the understanding of our world. Such creative piecing together of ideas of unknown provenance is also the source of useful metaphors, hypotheses and also underlie the socially important interactions of storytelling and humor. Nevertheless, actions based on unacknowledged or wignorant fantasy are potentially very dangerous.

5.15 Consciousness – An Analogue Process

Part of the definition of knowledge invokes the concept of consciousness. The exact mechanism of being conscious is not well understood, but it is not an empty concept. It is quite clear that consciousness is a process occurring in real time. It is also a prerequisite for the laying down of long-term memory and necessary for the activation of emotional circuits and the performance of deduction and induction. But it is not a binary function like the firing of a neuron. A useful metaphor would be the system of a dimmer switch and a light bulb. The amount of electricity flowing through a bulb and activating it as a source of light, can be anything from zero (full darkness = full unconsciousness) to the maximum current available (full brightness = full consciousness). This helps in comparing full consciousness with lower levels of consciousness, such as seen in day dreaming, incomplete anesthesia, and some posttraumatic brain states and so on. The nuances of brain function based on such a continuum of awareness will be discussed later in § 7.5. The point at this time is that there can be no knowledge without a brain-mind being in a conscious state, at least part of the time.

5.16 What Can We Know?

The above section begins to answer what it is to know. The content, however, depends on what reality consists of, what the world is in itself. At this point it is important to remember that the ontology of the universe includes not just matter/energy in time and space, but the relationships between the constituents. And for the purposes of philosophy, the relationship of the brain-mind to an external environment is of unique importance.

The processes of abstraction and imagination, as delineated above, can be functionally separated into the activities of the brain-mind as subsumed under such terms as anatomy, biochemistry, physiology, and the other areas of study necessary for understanding life. The ontological rules that govern the lower levels of being cannot be ignored, but the relationships of these constituents, as they give rise to the emergent possibility of abstraction and imagination, will need further discussion. Psychology, sociology, anthropology, history — all study the interactions of the output of brain-minds as they affect the internal and external environment. The resulting types of emergent processes define each area of interest. In the following discussion recall that the processes of abstraction and imagination both have unconscious and conscious aspects. By definition the unconscious parts can never rise directly into the conscious realm, and the conscious parts can influence, to some degree, such abstractions as categorization. In a feedback mechanism these systems also help to create a goal as one input for further imagination. In between these two are the hazy areas of preconscious and subconscious activity. We are not directly aware of the latter activities which do however have the potential to be the immediate precursors of those events that constitute the contents of our awareness. Some of the important preconscious abilities are the retrieval of concepts from long term memory, the emotions, and the latent recollection of the content of dreams that occurred just before the dim awareness of partial wakefulness.

Exactly what mechanisms are used to combine the building blocks, gained by the abstractions from external reality combined with the abstractions that are the models in memory, are hinted at by the products of the imagination. The paring down, categorizing and bundling of information in the unconscious abstraction processes add to the difficulty of assessing the source of some of the stimuli. This leads to several conclusions concerning the relationship between what we know and reality, the basis of material truth.

Chapter VI
TRUTH — and "The Thing in Itself"

6.1 The Kinds of Truths
One conclusion that follows from the Kantian concept is — what we know is not the-thing-in itself. It cannot be. This concept forms part of hypognosticism. From this it follows that what we call truth in everyday speech is not the same as reality. I denote that idea of baseline reality by the use of an all-capitalized TRUTH. When speaking, the phrase "baseline" or "absolute truth" could be used, but in writing the capitalized version is more graphic. In speech I prefer "baseline truth", because the alternate phrase is often used as a means of emphasis rather than as a clearly delineated subspecies of the genus name "truth". TRUTH refers to all of reality, it's components, relationships and potentials. All of which we can only partially know. The other subspecies of truth I categorize as follows.

Conceptual truth is akin to "knowledge of". It is partial; it is believed as a concept without internal contradiction within a total, personal worldview. When shared by the most informed specialized group on the subject, one of its word-symbols is theory. A subgroup is the concept of material truth which denotes the models based on the sensations.

Communal truth, namely the view held by a majority of a specified group, is more akin to the scientific concept of hypothesis. It is a model built on a (sometimes) acknowledged, limited set of data and points to some disagreement with those outside the group. In any given circumstance it can incorporate the whole range of agreement from just below the conceptual truth, to "common sense", to fantasy.

Neither of these has a direct bearing on logical truth which refers to the method used to arrive at a consilient concept. There are two subsets of this form of evaluating the truth, each forming their own field of study: the truth of a sentence which is called propositional logic, and mathematics. Logical truths are statements of relationships, independent of the material contents of the interacting things. This is why logical statements can be made independently of specific things, such as in: if X then Y. The most commonly given example is one of the variations of the following statements using predicate logic in the form of a syllogism:

(If) All men are mortal

(And if) Socrates is a man

(Then) Socrates is mortal.

The notations of logic share many commonalities with mathematical notations. Both are based on an assumption of internal consilience. Logical analysis concerning the real world gives conceptually "true" results only if the relationships are allowed by the P-laws that govern cause and effect. Mathematical systems use their own rules concerning the interaction of the symbols and concepts. Logical and mathematical equations and notations are subsets of the conceptual truth because they are based on the agreed to laws and axioms that are held in common by those who specialize in them. When these notations are used in philosophical writings, it is to summarize and formalize the flow of thought. I rarely chose use to them in this book because many readers may find them off-putting and confusing rather than a source of clarity. This parallels the method of writing about science for the nonscientist where chapter, paragraph and sentence structure are used. This is less efficient than equations, but more accessible, and it serves the purposes of general communication.

We all use loosely logical constructions in our everyday speech and writing. These statements often are of the predicate logic type, with an unspoken "given that", the equivalent of "if". In scientific and logic-based philosophical writings, the expectation is that the statements are, or can be, supported by the evidence, which ideally is both consilient and consistent. Such a consistent series of logical steps used to solve a problem is called an algorithm. However, even if the underlying logic is faultless, if the unspoken primary suppositions are wrong, we end up with erroneous results. This state of affairs is the point of the well-known phrase in referring to computer calculations, "garbage in, garbage out ", and can be applied to the calculations of the mind as well. The most accurate conceptual models are therefore the product of the faultless logic and the most reliable understanding of the underlying TRUTH, including the emergent properties found in the studies concerning subatomic physics through mind.

6.2 Finding conceptual truth

The process of finding conceptual truths in all humanistic endeavors, including, philosophy, must therefore follow methods comparable to science in the constant search for both consilience and consistency. Unfortunately, in the inquiries which study these high-level processes the experimental approach is often impossible. In some cases, this is due to the lack of the appropriate natural sensors or the instruments that extend their reach. This particular inadequacy has lessened over time and given us better data. The other reason such a process is usually not possible, is that the necessary control cases cannot be arranged. This can also be true in the physical sciences. In astronomy and subatomic physics, it is because of distances and sizes; in environmental studies (ecology) it is because of the inherent (and underrated by the lay community) complexity of the interactions. The study of mind shares all these problems. However, experimentation is not the only way to proceed in the search for more accurate conceptual truths. Good data can be gleaned from reality in another way, namely by the use of structured observation.

Structured observations come in two major varieties. The first starts with data from previous observations. The results are then categorized based on this initial set of raw data that has been observed and noted. The structure and function of our brain-mind initiates that process, even at the abstraction level, by applying preliminary categories to the incoming information. In vision this includes such characteristics as color and form. In hearing, an infant's mind already differentiates speech sounds from other noises. Touch input is unconsciously categorized as pleasant, noxious or neutral, before it is presented to consciousness. Consciously structured observations are based on similar information as the unconscious ones, but are modified by comparisons with similar previous observations stored in memory.

The second method for observational data acquisition, is more active. It is a guided search for data that is intended to fill in gaps of information and to increase the variety of sources. This is a conscious, goal-oriented search. This is what scientists such as Darwin did, at home and in his travels, by observing available environments. He was also inspired by the great Alexander von Humboldt. Both accumulated large amounts of data not previously available and added it to current information. In Darwin's case, some of the organizational concepts, such as what makes a bird a finch, had already been supplied by previous observations and categorizations, leading to definitions. But he noted some consistent differences in the physical appearance in this family of birds, from island to island, in the Galapagos. This consistent local variation in the cases of the finches, and many other families of plants and animals, then became data for improving the relatively new hypothesis of evolution.

We now have even more data, at different and deeper levels of observation, such as the genetic composition of a particular type of organism. This has caused the theory of evolution to be modified and improved. This well know example underscores the fact that any current conceptual truth is not the absolute truth for which so much of humanity yearns. This recurrent possibility of modification of a theory, by the use of new data, is used by some as an argument against the validity of our current understanding. This argument is based on a falsehood, namely that a tautology between the best conceptual truth and baseline truth (TRUTH) can be achieved. It is false because all our knowledge, our conscious concepts, are demonstrably based on partial data, no matter the total quantity.

6.3 Received Truth

A received truth is a truth based on transference of information from another sentient being. Even if an authoritative transcendent source of information (gods, spirits, alternate universe aliens) imparted a "received truth", this information is bound to be incomplete because of the limits of our brain -mind (unless one equates oneself to an omniscient authority). In any event, equating personal understanding with the full understanding attributed to such a source, is denied by most spiritual leaders. Even if such a transfer of information was the case, some energy would have to be expended to change the content, material, structure and/or energy flow of a brain. This goes against the P-laws that the matter/energy content of the universe as a whole remains the same throughout its ongoing changes over time. One could propose a change in this P-law so that new matter/energy input is allowed, making it possible for a transcendent cause to have an effect in this universe. But if this were so, all attempts to understand the workings of this universe, including life and mind are guaranteed to be undeterminable. Unknowable inputs and rules must lead to incalculable results that can only be self-consistent by the use of untestable, confabulated concepts.

Both the pursuit of scientific and modern, realist philosophical understanding is based on the belief that the universe does have a constancy of cause-and-effect relationships, unaffected by an outside, transcendent influence. Even in some sub-atomic theories, when the cause-effect relationship is questioned by the mathematical models, there are posited governing relationships. I would argue that our lack of understanding of these sub-atomic relationships and their actualization, does not preclude a different, particle physics specific cause and effect ontology that is the precursor of the emergence of the atomic world.

Theologies and other types of transcendentalism make assumptions that are all equally unfalsifiable and therefore closed to correction. Therefore, these revealed received truths are unresolvable, incompatible pseudo-conceptual truths with no justifiable connection to TRUTH. These authorities of revelation cannot even come to agreement about what counts as a conceptual truth, both among themselves or with non-transcendentalists (realists). Because of this, realist philosophers must find their own way to an understanding of the human condition without referral to any transcendentally posited influence on our world. Discussions with theologians and spiritualist are limited by the different understandings of the structure of epistemology. Both groups could, however, agree that there are mysteries, and aspire to hypognostic wisdom.

There can also be a transfer of one human's purported conceptual truth to another human, for whom it would be a received truth. In this case whether such information should be treated as knowledge or belief is, at least in principle, testable. This is further discussed in § 6.7.

6.4 A Return to a Defendable Quest for Understanding

The theory of evolution is, as all theories of science are, a portion of the best conceptualizations of the TRUTH at any point in time. All theories of science are in the international public domain. Although accessible to all, it usually takes deep specialization to create useful data and the fullest understanding. Basic accessibility is critical to the philosophical endeavor, because, philosophy needs to access the current fundamental, scientifically-based concepts to find a path to a consilient worldview.

Hypotheses and everyday truths, the communal truths, are less formal, less specific and more likely to be found to contain errors once more complete data and an internally consistent logic is applied. However, these kinds of concepts are useful, like instincts and emotions, in leading one's life. Communal truths, like theories, are obviously always open to improvement.

Non-fiction fantasies are not open to improvement because the believers of the fantasies do not admit that the data they are based on is incomplete, counterfactual or unsupported. Conceptually, that is the difference between fantasy, masquerading as a form of truth, and hypothesis. The latter acknowledges the hypognostic state, and therefore can be built upon and improved. Perhaps this is how the Socratic phrase, "the unexamined life is not worth living", could be interpreted. In other words, "don't lead a life based on fantasy", a conclusion I believe Kant might have reached if he had written a work called "Critique of Pure Fantasy".

6.5 The Limits of Knowledge: The Case for Hypognosticism

What cannot be known? Firstly, one cannot have knowledge of something that does not exist as an entity, relationship or process. Also, as noted above, one cannot have knowledge of that which is not part of this universe, that is, the transcendent. The latter may be a special case of the former because one cannot insist on the nonexistence of this often-proposed entity. I only deny an active relationship of transcendent entities with this, our, universe. This is the first part of the meaning of hypognosticism (recall § 1.1) Theological agnosticism is more specific, denying the knowability of a being with the attributes of a god. Neither is the same as atheism, or its equivalent unwarranted secular denial. Hypognosticism limits itself to this universe, and its direct predecessor, the origin of the "Big Bang", a lay-term for the sudden expansion of this universe from an unknown and undescribed (indescribable?) "singularity". The term indicates an appreciation for the fact that some complexities, in their totality, cannot be appreciated by the human mind. At best we can only know, at any one time, parts of reality. We can only understand details of its content, consequences and relationships, piece by piece. Human knowledge, that is the totality of all that has ever been known by any human mind, can only be sampled by any one mind a little at a time. Therefore, any general statement about reality, is most likely going to be less precise than a statement about a specific smaller piece. But as long as a general statement is consistent with the understanding of the smaller pieces, it is not only useful, but necessary, to navigate through the mazelike path of one's existence. Together, one can define hypognosticism as follows:

"The positive belief that any concepts concerning the contents and relationships in our universe are inherently incomplete, and that any concept of the transcendent is not a conceptual truth."

It is a humble yet hopeful philosophy, one that by its acknowledged incompleteness allows for progress and improvement in its content.

6.6 Truth and the Lay Uses of the Verb "To Know"

In common usage, the verb to know has many uses that go beyond conscious modeling of reality. Although all usages imply the need for consciousness as an underlying function, they do not meet the basic, narrower epistemological definition. Because the meanings are so often confused or misapplied, several are listed here both to clarify and delineate the differences in meaning of this polysemous word.

 A. — In a simile

> "A tree knows that winter is coming and gets ready to lose its leaves."

If the statement read "The tree "acts as if it" knows ...." It would be a metaphor. Attributing goals to non-sentient beings is a common fallacy. However, using the statement as a metaphor, can be a very useful communication device. As a simile, the sentence could be understood as a statement of fact. Whether the omission of the "as if" is done out of ignorance, mystical beliefs or just plain carelessness, this kind of use should not be found in a discussion of science and philosophy. This kind of laxity is also found in describing instincts and emotions in both anthropomorphisms and in these certain kinds of animal reactions.

For example:

> "A bird knows how to build a nest" or "Infants knows how to suck on a bottle"

B. — As a habit.

"I know how to get there without thinking about it."

If one drives or walks the same way to work every day, the turns we make, the routes we take, are followed in a semiautomatic way. It does not have to include the act of consciously thinking about it. That is not how we would handle the task of giving someone the information to follow that route. The two mental processes are obviously related but not the same, and only the latter fits the basic definition of a conceptual truth. The former use may be termed "motor knowledge" for differentiation.

C. — As a generalization.

"I know that my wife would like this song because she likes country music."

Here the verb is used to assert a belief based on a generalization. Unless she has specifically told you that she likes this particular song, you are making at best an educated guess. It is an example of the common error of misapplying predicate logic. If one were to ask her whether she really likes all country music, without exception, the considered answer would almost surely be no.

D. — As a cover for ignorance. –

"I know how these things work!"

In a sense, this is an excessive generalization, but in this case, it may be driven by an unconscious wish not to seem ignorant about a portion of reality. This cover, to protect one's own self-esteem, may be conscious — in which case it is a lie. If the untruth is actually believed, it is a form of self-deception. The latter is different from an ordinary generalization in that, if questioned about that conclusion, the same phrase would be asserted, perhaps indicating a subset of the claimed knowledge.

E. — As willful ignorance.

"It is known that the earth is only 6000 years old."

This is an assertion based on ignoring the best information available, and is especially pernicious when used in the public sphere, such as politics and education. This kind of assertion is so commonly used in American public discourse that it deserves the aforementioned (§ 5.13) neologism, wignorance.

(The term has been used as tags in Twitter and Facebook, but the context is not clear). There seem to be two major causes for wignorance. The first cause is a belief on the interaction of the transcendent with the reality of this universe, thereby denying the P-laws. As noted, before, such use allows one to believe or "know" anything at all. Such unbounded inclusiveness cannot lead to more accurate models except by happenstance.

The second cause of wignorance is often found, in cases where the conclusions, based on well researched data and faultless logic, go against what one would like to believe. One may prefer a simpler or less disturbing worldview. This emotional interaction with cognitive thought is very common to all of us, but should be consciously searched for and avoided when the goal is knowledge and wisdom.

F. — As used with acknowledged ignorance.

"I know there are approximately 160 countries in the world right now."

This is an acceptable use of the basic verb. The use of the modifier indicates that one is aware that information is incomplete, but nevertheless based on reality. It is an acknowledgment of the hypognostic worldview, and such modifiers should be used more often. In some quarters, however, this is seen as a form of stupidity, or unwarranted ignorance, when in reality it indicates the humility of acknowledging the limits of memory and the recollections based thereon. Academic one-upmanship is a common source of such putdowns. It is an attitude and source of accusation which do nothing to further the search for mutual understanding.

G. — As assertion of a definition.

"I know his name is John" and "I know that is a cat."

Such phrases are usually given in the shorter form of "This is John" or "This is a cat". It indicates that one believes that a particular word or phrase refers to a particular concept in the linguistic community one is a part of. It indicates that the word is a tautology for the concept. When the verb "to know" is used in such a case it is usually meant to emphasize the correctness of the definition.

In the normal flow of speech, one is usually not consciously thinking about what the next word will be. The subconscious threading together, from memory, of words and phrases into sentences and discussions is awe-inspiring.

We are especially cognizant of what we have said if we realize that our words, that we heard ourselves utter, are incorrect. Or that the words may be misinterpreted (as noted by the reaction or response of the listener). We then, subconsciously, search for a more appropriate phrase, and consciously evaluate any that come to mind, and then know (in this form of its use) that it transmits the concept we are trying to transmit.

H. — In describing oneself.

"I know that I: - love you; -am confused; -am skillful; etc.".

This can be seen as a special case of type" G", but for a set of concepts of the speaker that cannot be verified by others.

One of the Delphic maxims is "Know Thyself." The difficulty of this attempt is emphasized by the common idea that "Others can know you better than you know yourself". This brings up the question of whether a mind can truly know (have a consilient and correct model) its own outputs. This would mean having a conscious model of the modeling organ and its activities. I believe it is possible, to some degree, by the same method that scientific knowledge can be gathered when experiments cannot be performed. But this self-knowledge cannot reach the same levels of certitude because independent verification of our subjective feelings and many memories is not possible. Only our actions and their results can be noted and studied by others. Modern science has however increased the kind of things that can be noted and has made possible the application of new experimental methods to the study of the mechanisms that lead to our actions. The use of functional MRI in researching the unseen "private" activity of the brain-mind is now a common method of investigation. Experimental psychology makes use of the manipulation of the environment of a subject to study the resulting "public" actions. Both methods are yet in their infancy, but enough data has been garnered that tentative conclusion based on consilient interpretations are now possible. Such interpretations have, and will continue to have, a major impact on progress in philosophy.

6.7 Can Knowledge Be Received?

No one human mind has access to more than a tiny fraction of all data available to humanity, as a group, over space and time. This gives rise to the concept of special knowledge that is held by an authority, rather than knowledge based on personal experience and imagination. But knowledge (the exact mental model) cannot be transferred, only information can. This information, once processed by abstraction and imagination in a receiving brain-mind and returned to its conscious component, can then be referred to as knowledge. This indirect transfer of a conscious concept from one person to another, can lead to agreement, disagreement or puzzlement. It is one of the main reasons why teaching does not lead directly to learning, either in the classroom or in life. This process also emphasizes that different human minds have a unique history in turning data into models. So, the answer to this Section's question is no. What is received is information. When the purveyor of the information is a teacher, leader, guide or a role model then we may place that person in a position of authority and perceive the underlying information as a source of truthful knowledge.

There are two main kinds of authority; the specialist and the generalist. The specialist concentrates on one specific set of data in depth. The generalist takes the concepts, as created and agreed to within the different specializations, and then integrates these concepts to form new concepts that include data from a larger section of reality. One could also call a generalist a meta-specialist, to parallel the idea of meta-analysis. In meta-analysis data from many different but related experiments are combined in an effort to create more accurate answers to a question, and to find consistency and then consilience in the larger data set. Specialists are best at finding detail and differences. Generalists are better at finding connections between concepts that may seem to be contradictory, or even unconnected. The resulting combination of concepts is critical in understanding the bigger picture. Scientists are more likely to be specialists, but may be generalist. To some degree the opposite is true for philosophers. Both kinds of minds are needed to move understanding forward in any field.

Agreement between authorities, does not guarantee the best interpretation, and may even be dead wrong. The reasons for this state of affairs may not be either form of ignorance. Assuming that the data is not tainted, the problem is often due to poor logic. Another major reason for coming to wrong conclusions is found when a better answer is not sought, but rather the data is shoehorned into a preformed concept. This latter reason is commonly associated with heated arguments and the animosity engendered when the preferred concept is closely associated with the ego. The data may of

course also be an error, due to incorrect measurements or faulty collection practices. This tendency of setting aside of or ignoring the data that does not fit the current model, is another form of wignorance. Scientists and philosophers are only slightly less susceptible to wignorance than the typical human being, but because of their position in society, should be held more accountable.
The errors leading from wignorance are known since antiquity.

"True wisdom is knowing what you don't know." –Confucius

"The only true wisdom is in knowing you know nothing." – Socrates

Both of these sayings only make sense if the phrases indicate a sense of ignorance. A literal interpretation would make it impossible to discuss anything at all. "Knowing what you do not know" can only acknowledge the edge of what is known.

To "know nothing" can only be interpreted as saying either that the concept "nothing" refers to a thing in reality, or instead, more generously, that the knowing is an incomplete model of that reality. An alternate, literal, reading could indicate that knowledge is not about reality, but is a total fantasy with no connection to reality whatsoever. This would posit the possibility of a lawful energy transfer between something outside the brain that is meant to deceive, as described by Descartes as a malevolent demon. As noted, before, such ideas of correlation, unconnected to reality, are fruitless and create a hopeless pathway toward understanding. Such ideas do not point to ignorance of the yet unknown, but to fantasy.

Therefore, the concept of knowledge is most fruitful when it is seen as an incomplete model with a connection to reality. Metaphorically conceived of as an imperfect, abstracting mirror. In that case leaning on the authority of only the specialist or the generalist brings specific problems to the fore. The danger of over-specialization is that it may consists only of the process of accumulating data, and that its interpretations are not asked to lead to an overall consilient or consistent world view. The danger of generalization is that it may give the wrong weight to any or all of the components. Progress is more often found in combining the two methods. The specialists do their best work when they look at those areas of current data that are incomplete, or seem to contradict the best generalization. The generalists do their best work when they constantly update a concept as newly generated or discovered information becomes available. Therefore, the typical human should be very wary of any absolute statements concerning concepts generated by authority. The level of incompleteness of knowledge can be measured by the number of questions that cannot be answered by the generated models. It is best to keep a hypognostic frame of mind. In this context perhaps the word pauci-gnostic would be more appropriate. Nevertheless, received knowledge from authorities, although often representing the best current data and models, should be seen as a way-post to future knowledge, not as a final destination.

6.8 What to Do When Authorities Disagree
Disagreement can be at the level of data or the concepts derived therefrom. Specialist can speak best to the weaknesses inherent in the gathering of data since that is what they do. Generalist can clarify the situation by attempting to fit together alternative concepts within a field by judging which concepts are more consistent with the concepts of adjacent fields, that is, in a larger system.

If a particular concept is totally inconsistent with an accepted consilient picture, there are two main ways forward. One is the abandonment of a group of data. This does not necessarily mean that the isolated concept is the one that needs to be abandoned. Comparison of that inconsistent concept must also be made with the sub-concepts of the consilient picture. This is what imagination does best, since it only involves the energy and circuits needed by any mind to function. The second method is to gather further data that would test the concepts involved. This takes a lot more effort, since it involves not just a mind, but manipulating the world. But both are needed for progress to occur in the long run. The typical person has neither the time available (s/he is too busy meeting life's basic needs) nor the intellectually inquisitive inclination to do more than pick one authority over the other. But the abandonment of a consideration of alternative authorities is a dead-end in the search for wisdom. Making a final conclusion concerning a controversial topic, by avoiding new information and then never revisited, is also a form of wignorance.

Today, the choice that is most fraught, is the one between the authority of religion and science — both seen in the broadest terms. Whichever authority one chooses to follow can cause one to magnify the disagreements with the other. True progress in understanding will only be made, if the leaders of the various authoritative branches of understanding put aside wignorance and acknowledge the inherent ignorance of all camps. Wignorance is the greater fault of many on the religious side. An unwarranted sense of certainty, based on shared human experience, as filtered by logic, is the greater blindness on the other. Perhaps a sign, placed on the wall across from the working desks, stating that "Ignorance is inevitable" would lead to a more fruitful conversation. This phrase could be the watchword for hypognosticism.

6.9 Culture and the Tribe as Authority

Which choices of authority one is aware of and considers seriously, is highly weighted in favor of the consensus within a culture (Greene,2013). This is especially true if the culture is monolithic as it is in smaller systems such as tribes or clans. A tribe can be classified as a genetically related group — and that is what it was when distances or physical borders isolated groups of people. As the world became more populated, tribes cooperated — or fought over resources and ideologies. The latter, questioning of beliefs arising from the clash of concepts, is the heart of the remainder of the book. What is a right versus the wrong way to achieve any end? What should that end be? Which gods or authorities, if any, should be worshiped or held in esteem? Justice, ethics, purpose, the meanings of emotions, are all subheadings of those questions that pertain to the evaluations of relationships. These concepts bring to the fore the importance of deciding which authority to follow on any group of issues. Authorities are needed. The individual cannot even begin to create a cerebral, versus an instinctive, worldview based on more than a tiny fraction of all the pertinent information available to humanity as a whole.

Thus, the place of the philosopher as the premier generalist of knowledge (albeit s/he usually is, especially within academia, a subspecialist). It is my philosophical view that the uncertainties of knowledge are not emphasized enough. The differences between isms are highlighted, but those that have been found to be untenable are not deleted from the syllabus. In science this would be like continuing to have chapters on alchemy or spontaneous generation of life from rotting meat.

Philosophical studies require a stable environment, without frequent threats to one's own personal existence. Full time, professional, philosophers are therefore privileged persons, that do not need to spend large portions of their efforts on the tasks of finding food, shelter and the other necessities of life. Priests and shamans were perhaps the first — they were not expected to mainly hunt or fight or gather food. Their needs of life were supplied to them by the rest of the tribe. The other tribal members exchanged the material fruits of their efforts in return for the wisdom that can be the fruit of prolonged thought and the gathering of wider experience. But, like the neuron, such a proto-philosopher can become a bottleneck in the formation of concepts, bringing together many inputs, creating one conclusion, and then, with unwarranted certainty, distribute just that one limited result.

Chapter VII
Concepts

7.1 The development of a concept

Concepts are developed or adopted on two levels — personal and tribal. The dynamics of both are rather different.

The personally developed concept is highly influenced by the particular genetic, neurodevelopmental and macro- environmental settings of each individual. It allows for more variation, creativity and perplexity, than "groupthink", the concepts held by a group. A personal philosophy is shaped by the forces centered on the self and gives rise to a personal vision of an ideal world — an egotopia. In this book, the word is meant to contrast with a utopia, an ideal that is shared by a group of similar minded individuals. In passing, it should be noted that utopian communities rarely last more than a generation, and most commonly disappear in a span of a few years. The dissolution of utopias comes from the death of the leader, or contentions for leadership based on similar but different personal ideals. Egotopias are often situational and very short-lived as new information impinges on the holder of an idealized concept. However, such ideals are the basis of the individual concepts of justice and ethics as well as beauty and evil. The discussion of these topics will be amplified in the third part of this book, but all are concepts and not things that are independent of a mind.

Although some philosophers lead hermit-like intellectual lives, most have some intellectual relationship with another person or a small group. The development of concepts under these conditions are amendable to investigation. One could call the former a psychologically based philosophy and the latter a sociologically based philosophy. In the end, each kind of philosophy interacts with the other type, but, for discussion, one can usefully separate them.

The individual is the only source of creating new ideas. Those ideas can then be reshaped by interaction with other minds, but the seed of each comes from an individual. Wisdom is a result of such creativity, where one person "sees" new or better concepts by the application of the imagination to the data available to that particular mind. But to grow, such wisdom must be shared, compared and then reworked.

7.2 Ontology of a Concept - the Basis of Epistemology

The detailed discussions of the ontology of a concept have been left to this section because it is an integral part of epistemology. The emergent possibilities of mind that create concepts are so far removed from the emergent possibilities of the lower levels of relationships and structures, that they are best examined here.

Mental models can be divided by the source of their most proximate stimuli. The models concerning the material world, as abstracted by our sensory system, are relatively straightforward. They can be compared and tested against similar models of other minds since all refer to the same or to very similar physical sources or phenomena. The comparison of models of human relationships is less easily done. The relationships between material/energy things can usually be described by referral to agreed-upon metrics of distance, volume, time, mass, charge, etc. Comparative concepts such as far and near, large or small, hot or cold or a sound that is loud or soft can all be measured by some standard metric. But what is far for a snail is less so for a mouse, never mind a bird. Nevertheless, with a small amount of context, these relative descriptive words for their underlying concepts can be understood appropriately.

The concepts for psychologic and social relationships are much more difficult to convey and have filled multiple shelves of innumerable bookcases with tomes on these subjects. The concepts of thought and mind are based on and include all the previously mentioned emergent possibilities, but they are not as open to investigation and their metrics are much more difficult to devise and measure. New insights and tools are needed. These concepts are garnered from studies in: neurology, both anatomic and physiologic; psychology, experimental and observational; and the social sciences. All these fields of study are engaged in the ongoing tasks needed to advance toward the goal of understanding the brain-mind and its place in human relationships.

The underlying CNS networks composed of nerves, and their interactions, are being studied more intensively, creatively and thoroughly than they ever could be before. However, the emergent concepts of justice, mercy, evil, prejudice, beauty, Schadenfreude, pity, and many others could not be measured or studied using standard physical measurements. They could only be measured, in the past, by comparing self-reported introspection, behavioral consequences and inferred intentions. Physical studies, such as the use of functional MRI can now further clarify some questions. These in vivo physical studies can provide an answer as to which combination of circuits are involved when a test subject reports a particular mental activity. Secondly, they can provide comparative data when different subjects report the subjective interpretation concerning the same objective stimulus. However, these studies have also shown that there are, as is said, many ways to skin a cat. This means that the general picture of mental events is the same between subjects, but always subtle variations occur between subjects, and even within the same subject from time to time. This should not be surprising because the brain-mind is such a unique and complex combination of structured matter and organized energy flows in both temporal and spatial relationships.

A as a metaphor, I offer the following. If one organizes multiple marathons, over the same course, involving the same 300 participants, the results would be slightly different each time. We would certainly note that the results would have a certain pattern overall, but the details would be different, sometimes surprisingly so. Those surprising outliers from the general recurrent results (say from 100 races) might be inexplicable except by noting small incidents particular to each race. The outcome of the races is not chaotic or random, there is a pattern to the results, but the results expected from any group of races can only be described in a probabilistic way. It seems foolish to expect more consistency from a brain-mind interacting with a more complex environment. But such foolish expectations do not seem to influence our psychological need for precise, unrealistically precise, predictions and answers. Part of wisdom is knowing that knowledge is always incomplete. However, part of wisdom is also knowing that understanding can always be, and should be, improved. The need for acceptance of this kind of wisdom is summarized by the popular quotation from the writings of the theologian, ethicist and commentator K.P. Reinhold Niebuhr. The "Serenity Prayer" is quoted in Wikipedia as:

"God, grant me the serenity to accept the things I cannot change, Courage to change the things I can, and wisdom to know the difference."

A secular variant could easily be:
"Accept with serenity the things that cannot be changed, have the Courage to change the things which should be changed, and the wisdom to distinguish the one from the other."

Although the original is in the form of a prayer to a God, it shows how quite different world views can still show some agreement.

Progress in understanding is greatly hindered by a lack of agreement about some of the basic properties of the working mind. The first property is that all concepts cannot exist without an active brain–mind. The habit of attributing this level of complexity, in some cases to even nonliving entities, just muddies the waters of understanding. Although using more basic systems, as a source for metaphors, in describing more complex ones can be helpful in understanding, the reverse is not true. This is because the emergent properties of more complex systems do not, and cannot, exist at lower levels of complexity.

The chapters on justice, free will, good and evil, and so forth, will add much detail to this preliminary outline. Here I will discuss what they have in common, as subjects of knowledge. To start with, are concepts such as justice, based on a real material/energy thing? I propose that they are, but not in a typical way. All these evaluative terms refer to interactions between a brain-mind and material entities, especially other brain-minds, or between different subsystems of the same brain. Note that the source of the concept, brain-mind, is itself an atypical thing. What is involved in the creation of all these concepts is the flow of energy (mind) within a particular brain, the combination called brain-mind. These active concepts can then act as stimuli for a feedback loop or physical action. This epistemology is consistent with the modern form of conceptualism (McDowell, 1996).

7.3 The Energies of Mysticism

The source of the mystical ideas concerning the flow of unknown energies between external concrete things and a system such as the brain is, perhaps surprisingly, partially right. Although the details of the typical explanations are complex confabulations, conjured to explain observable relationships that have no obvious cause and effect interactions, they can be based on real physical energies. In the effort to help clarify this, I offer this example from personal experience.

Several years ago, I participated in a group activity called "Build your own Theology" in a local congregation. This endeavor is supported by guidebooks published for this use by the Unitarian Universalist Association (UUA, 2021). This activity would more accurately be called "Build your own Philosophy" because a belief in a god or other transcendent beings is optional for most UU congregations. (SEE Glossary for the seven principles).

In one exercise we were paired up, and then each person would try to explain to the other, without five-minute interruption, some of their beliefs. This would be followed by a more interactive give-and-take of ideas. My partner in this endeavor tried to express her feelings of empathy by talking about receiving "energy" from me — sensed somehow in her abdomen. Science has no information about such undiscovered energy, particles or waves that transmit empathy, and/or receptors for such energy forms that then connect to the afferent nervous system. The was important for me to gain an understanding of her message, rather than to respond with secular disdain or full disbelief. I explained my interpretation of her feelings of energy transfer in the following way. My response, concerning her feelings of energy transfer, was the consideration that perceived psychological interactions could have a cogent alternative explanation based on scientific data. If that is not possible, then one must acknowledge a lack of pertinent information. The following is how I implemented hypognostic ideas in this personal interaction.

I started with some basic information. The first facts are that we have and say there is such a thing as "body language", as originally described by Charles Darwin. This form of communication can be shown to be the result of perceptible muscle movements, large and small, of the trunk, the extremities, and especially the face. The muscles that control the complexities of speech are also a major actor. It has been clearly shown that these actions are registered, abstracted and passed on into the deeper nervous system where they can trigger the activity of the so-called "mirror cells", or better "mirror system". The details of these systems are still not clear, but one cannot deny the recent functional MRI findings of subconscious activation, in well-defined areas of the brain, that corresponds to activation during similar self-initiated actions. The meaning, that is, the consequences of that activation, is more controversial. For a short, readable discussion see the interview with Neuroscientist V.S. Ramachandran: (http://greatergood.berkeley.edu/article/item/do_mirror_neurons_give_em pathy), 2012

The controversy is not whether or not there are downstream consequences of these subconsciously registered gross and fine muscle movements, but rather which and how many systems are stimulated to a greater or lesser degree. Because the nervous system is a system of systems, with constant influence from both the internal and external environment, one must ask the appropriate questions. We have discussed how the hundreds of thousands of impulses received by a nerve result in the binary outcome of discharge or no discharge down its axon. Furthermore, although each nerve is capable of only responding one of two ways (as far as the influence on the down-network nerves are concerned), each tiny input has a consequence. The sum of excitations and inhibitions caused by incoming signals result in a constantly changing probability of activation of an actual impulse. The outcome is one or zero, depending on whether the nerve fired or not. Another question is, does such a small change in one nerve affect other nerves? The short answer is yes. But. and this is critical, one input alone is never the only factor, and cannot, by itself be called the cause. The causes of a discharge of a neuron are never the result of the chemicals released by just one particular adjacent nerve ending on the receiving dendrites or other parts of the receiving cell. It is not, and never was, a particular "straw that broke the camel's back". At most it was the straw that most recently contributed to the total load. Or, to use another metaphor, the snowflake that most recently added its tiny weight resulting in the destabilization of a huge mass of snow and "starting" an avalanche is not the cause. Even in such a scenario, it might instead have been the melting of another flake that caused the destabilization in an ever-changing dynamic system. This is why avalanches, volcanic eruptions and human responses are often very unpredictable in probabilistic terms. Never-the-less, probability can more precisely model the quickly changing dynamics of complex systems than simpler, one cause - one effect, models. Unfortunately, the human brain cannot calculate probabilities correctly in all but the simplest systems.

What do avalanches and camels have to do with a nervous system that creates empathy? They are all the outcome of uncountable dynamic atomic and molecular interactions, each of which lead to an all or none response. The size of the response can vary greatly, because both the mind and a massive snow accumulation are super-dynamic systems, in which smaller subsystems can also be affected, or not. This leads to a seemingly continuous, analog response pattern. At the base, however, they are always digital, that is quantal effects. An electron does, or does not, jump to another energy level in the atomic structure in response to an input or loss of energy. The nucleus of an atom does, or does not, fission, and only in a finite number of ways. And so it is, all the way up to our subconscious or conscious empathy toward another living being. The final effect is made up of innumerable individual all or none responses to all stimuli, perceived consciously or not. There are energies leading to empathy, but they are quite the normal kind.

7.4 What Do Plants, Animals, Infants and Children Know?

It is common to read anything, short of an academic or scientific paper, without running across multiple references to what plants, animals, embryos, etc. "know". That human infants and children, as well as many animals can create mental models of their environment is not in question. What is problematic is the assumption that the content and development of the mental model are the same as those of a typical adult human being. Furthermore, an atypical human brain-mind is exceptionally difficult to understand. Nevertheless, an attempt toward understanding the differences will be of high importance in discussing the concept in the following chapters.

Plants, which have no nervous system (NS), cannot, by definition, know, anticipate, realize, feel, or have any of the properties that require the emergent properties seen in a nervous system. The fantasy of an aware plant kingdom was common in ages when information concerning the properties of living cells were essentially unknown. It is the basis of animism and pantheism, classic examples of the power of the human imagination at work in creating fantasies. No matter what the limits of available information are, this imaginative process is capable of deriving explanatory concepts of real-world phenomena by the use of confabulation. The study of these low data models, being based on information now known to be drastically incomplete (and therefore open to greater error) is the subject of the history of concepts. However, relying on these historical models as a base for one's own beliefs and actions, namely one's philosophy, is to build sand-castles when more solid structures are needed and desired.

Animals do have a NS, but the total system complexity and total nerves per system is usually much smaller than in the typical human being. To stretch the above metaphor, most animals can only build castle outlines without rooms or contents.

Infants and children have the genetically endowed brain capability to build sturdier and more complex castles. The genetics provide the architectural plans, but the materials must come out of experience. As more materials are gathered, we see a constant restructuring as well as growth and strengthening of the possible mental abilities. This process continues well into adulthood. With time, the repair mechanisms needed to keep the structures intact decline, or whole sections are destroyed. To complete the metaphor, we will find a castle in various states of ruin. With death the physical structure is in effect vaporized and all the emergent properties of that brain-mind, including personal knowledge, are gone.

7.5 What is Consciousness?

Most multicellular animals have a nervous system. Those that do not can be lumped with the plants when discussing what it means to know. It is however critical to have an awareness of the relative complexity and actual structures of the different nervous systems in order to grasp what is meant by consciousness, a requirement for knowledge as briefly noted in the previous section.

Rather than define the phenomena of consciousness by how it emerges with increased nervous system quantity and quality, specific structures and complexity, I propose to discuss how it differs from other related nervous system functions. The only touchstone we have, however, is our own personal experience, and that reported by other human beings.

For the purposes of this section, the division of the human nervous system into the conscious, preconscious, subconscious, and unconscious parts is useful. These divisions, as defined, are basically functional rather than structural. A rough thumbnail sketch will outline the differences.

In Daniel Kahneman's, book "Thinking, Fast and Slow" (2011), based on his work with Amos Tversky, he divides brain functions into a precognitive System 1, versus the cognitive System 2. It was based on research concerning what biases affect decisions under many conditions. I have divided the preconscious further, as will be discussed in § 9.11 (including Table II" The

Evaluative Systems".) The emphasis is on the effects of the mind's subsets of functions, on philosophical discussion. Kahneman and Tversky made what are considered groundbreaking changes in how we approach the functions of the mind. These ideas led to a Nobel Prize in Economics (2002) for Kahneman after Tversky had died. Although the following "labels" are also used when discussing Sigmund Freud's ideas, they are not exactly the same. This is mainly due to a different emphasis and the passage of time.

Brain activity can be classified as follows. Unconscious activity never directly reaches conscious awareness— this includes the activity the autonomic nervous system, such as blood-pressure and body temperature control. We find this in all mammals, birds and some other phyla, although the mechanisms involved may differ. Preconscious activity underlies the next higher levels, and without it, consciousness would not be possible. Under this term I would include such activities as storage and retrieval of memory and the afferent abstraction cascade that gives rise to the emotions. The preconscious activity uses the unconscious as input leading to thought and imagination in a recursive way. It can also be triggered by conscious will. This includes purposeful recall from memory and goal driven thought. Together, these two interconnected systems are referred to the subconscious. Consciousness, by default, is the end result of these processes. It is further defined by a few attributes the others do not seem to have. The first is that conscious experiences are capable of being stored in recallable memory that may be accessed by the preconscious processes. The second is that consciousness is necessary for goal-oriented activity. To further clarify the differences let us revisit some of the nervous system activities that do not include or necessarily lead to consciousness.

Reflexes, instincts, and emotions: we are conscious of simple "knee jerk" reflexes, if at all, only after they have occurred. This differentiates them from mental activities which depend on some planning. Also, consciousness of reflex activity is always about the recent past. Like simple motor reflexes that do not require consciousness, instincts are the complex coordinated reflexes that respond to more complex environmental stimuli. If one avoids getting close to a source of potential harm, without prior experience and before there is damage, discomfort or pain, that could be an instinct. Broadly, this instinct is the complement to another instinctual behavior, exploration. The carrying out of the instincts will require some level of awareness if a significant passage of time is necessary completing the action. Emotions are also complex reflexes, but in this case with the added efferent outputs that stimulate both hormonal systems and autonomic nervous system functions, such as blood pressure and heart rate. Although the strongest emotions seem to be created while we are conscious, they can also be instigated by internal stimuli during REM sleep (See below) and subconscious activity. The latter results in what is called mood. What differentiates mood from emotion is that the former is usually of greater duration. Secondly, we are always aware of the latter, but need not be aware of the former. The two concepts otherwise overlap greatly and may flow into one another.

Habits: the relationship between physical habits and consciousness is similar in that it is an efferent activity. The main difference being the potential complexity of the action, and that it is learned, that is, previous experience and memory are involved. Playing a violin concerto "by heart" is such a complex activity. Mental habits are also based on the pathways of energy flow formed by previous repeated activity, but without a motor component. They are difficult to observe in oneself and only inferred in others. Learned prejudice is one of these preformed patterns of reaction.

Sleep: in sleep all three lower activities of the brain can be noted. It is also the one activity that makes us most aware of the preconscious processes that occur at the border between full awareness and a kind of observational state that is only loosely connected to conscious control. Sleep is divided into four stages by the American Academy of Sleep Medicine. They are REM (rapid eye movement) sleep and three non-REM stages, N1, N2, and N3. REM sleep is the lightest and is when we dream. N3 is the deepest, and at that level, the brain is least reactive to outside stimuli. The latter is the closest we come to general anesthesia and full unconsciousness without the use of drugs, trauma, disease or lack of oxygen and glucose. Indeed, that level of decreased consciousness does not occur in sleep. In the former cases, the altered consciousness are forms of brain dysfunction. The states between sleep and awareness are called hypnagogic and hypnopompic. The first is the transition of wakefulness → sleep, the latter sleep→ wakefulness. These mental states in the gray zone between wakefulness and sleep are of interest to philosophy as it involves discussions of dreams, memory and other mental phenomena.

Awareness: this is a state in which we claim to be conscious. It is closely associated with consciousness, but is only a subcategory thereof. In a Venn diagram it would be a smaller circle inside a larger one. It can be formulated as being conscious of being conscious – that is self-conscious – but without the emotional overtones. It allows us to create wanted memories and to react purposefully to our environment. Neither of these occur with the reflexes, deep sleep or in the unconscious state. Motor habits can be seen as a special intermediate case; in that it involves a special kind of memory. It is not the memory called up for thought, but rather something called "motor memory". It has parallels with automatic activities such as speech, which is also an efferent motor activity, but habit is a use of memory whose actions are disturbed, rather than improved, by thought. Any athlete or musician will tell you that if they consciously think about what they are going to do in performing their skills, the skills will be suboptimal. Habits are therefore somewhere between reflexes and conscious activity. None of the above is meant to indicate that there are clear delineations between these categories. There are gray zones at the intersections, but the typical activities can be clearly delineated.

Concept memory: this type of memory is both a result of and is needed for conscious activity in a feedback loop. Activities that are performed only from reflex, instinct, emotion or habit, and that do not rise into awareness, are not laid down in concept memory. Memory is a semi-permanent change in brain-mind structure and function based on previous strong or repeated stimuli. As such, it serves multiple higher functions of the nervous system, and is critical for all the topics of philosophy such as justice, evil, beauty, and so on.

Thinking: this is the result of integrating, by the imagination, short-term memory and recalled long-term concept memory. It is the conscious equivalent of dreaming, but with the potential of laying down memories. Unremembered dreams, those not arising just before consciousness, do not seem to give rise to memories. Whether they change brain structure that may affect mental habits is not known.

7.6 What is Wisdom?

Wisdom is the application of accurate models toward predicting outcomes of relationships and then acting on that basis. Whether something was truly wise can only be determined post hoc, but its success is improved if it is prospectively infused by experience. Wisdom should not imply certainty. If the rational goal of philosophy is the pursuit of wisdom and if the personal intensity of that search is measured on a scale of yearning for conceptual truth, then two things are needed. As a rational pursuit, wisdom requires a constant refinement of concepts. The emotional satisfaction comes from the assessment that the goal of better understanding is closer now than before. One of the basic characteristics of wisdom is that, in its pursuit, one seeks the proper level of relationships any particular concept can be applied to.

Wisdom requires rational thought. This is largely mediated by the use of words. Words are a verbal shorthand for things, relationships, ideas and other concepts. Words began as sounds useful for communication. They were not created and learned in asocial isolation, but are still the result of a communal environment in the formative stages of a brain-mind. Infancy and childhood are well known to be the most receptive time for the acquisition of the symbols used for communication. There is good reason to believe this is true because the typical brains of these age groups are still growing and are more easily molded by environmental input.

Wisdom implies a continued immersion into the flow of information between the self and the world — with subsequent reiterations of that flow within the brain-mind which leads to fine tuning and greater clarity. For commonly held concepts, the shorthand of words must exist for efficiency in thought and communication. A discussion of the creation of such a system of symbols of meaning will be part of the next chapter. The accuracy of the concepts concerning our more complex interactions with the world depend on the accuracy of our models of what is — ontology — and what the source and limits of those models are — epistemology.

Concepts concerning relationships do not deal with what matter and energy things are composed of as such, but with the changes that occur in response to the interaction of things. To discuss the relationships within social systems, a major concern of wisdom, concepts such as justice, freedom, evil, etc. are used. For further discussion it will be helpful to outline the potentials and properties of some material systems. I found the following three categories to be helpful.

NLT — non-living things
RLT — reactive living things
MLT — modeling living things

An outline of these terms and relationships are given next.

7.7 NLTs. RLTs and MLTs – Categories of Being

The content of the set described by the first term (NLT) is fairly clear-cut in that they do not meet the definition of a living thing (See § 3.3). This group is most helpful in that it provides a baseline. The next term, the RLTs, incorporates the properties of the NLTs, but with the addition of unique properties that emerge with life, with and without a simple nervous system. The level and type of reaction to stimuli is based on the emergence of ever more complex functions. The third, the MLTs, includes the modeling of the world. This new ability necessitates more complex mental functions such as memory and imagination, and leads to the capability to choose between the mental models. This last ability is what allows the evaluative functions, the content of Part III.

The boundary between the NLTs and the RLTs is much clearer and much easier to distinguish than the difference between the two living types. This is because in the latter case there is no clearly demonstrable cut off, but rather a continuous overlap of properties from genus to genus, from species to species of each genus, and even between individuals within each species. However, the difference is clear, for example, between an RLT such as a jellyfish and H. sapiens, our best studied MLT.

When concepts concerning any interaction are discussed, entity to entity, entity to group and group to group, the referral to these three levels will allow greater clarity. Human relationships involve all three tiers, both in its internal and external relationships: that is, self - self and self – other interactions.

-

Table I
Characteristics of the interactions of existents

1 * NLT↔NLT (always bidirectional)

These interactions can all be described, in principle, with basic concepts from physics and chemistry. They are of philosophical interest because they are the best understood interactions and because they demonstrate significant limits of the possible properties of the higher levels, which also depend on energy and matter interactions.

2a * RLT→NLT

The reactions of an NLT to an interaction with a living thing is totally circumscribed by the possibilities of an NLT – NLT interaction.

2b * NLT→RLT

This is the domain of such relatively simple cellular and organism reactions such as tropism, reflexes and instincts. The NLT supplies the stimulus which causes a predetermined (type 1, NLT↔NLT) cascade of effects to physical parts of the RLT including those that were not directly stimulated by the NLT. For example, in tropism, these cascades can be further subdivided, conceptually, into the functions of metabolism, the biochemistry of a cell, and changes in intra-cellular structures. In the case of reflexes, primitive emotions and instincts these cascades travel from cell to cell. This is also the point at which variations in structure may result in more efficient and life preserving cascades. This can be seen as one of the earliest forms of variation imparted by the process we call evolution.

3 * RLT↔RLT

This next level of complexity is similar to the 2b level, except that the stimuli traveling in both directions are the result of more complex cascades of physical and chemical effects. Because of the complexities there are often feedback loops that extend over time. This feedback continues until the RLTs separate, come to an equilibrium, or one (or both) of the RLTs drop into the NLT state. Despite the increased complexity, these interactions can be fairly well studied, described and predicted. This ability to model and understand these interactions is predicated on the fact that an RLT does not exercise choice — an ability that requires a brain-mind and consciousness.

4a * NLT→MLT

This is a step up in the potential complexity above those allowed in level 2(b) interactions. This is because the tropic, reflexive, instinctual and emotional reactions may now be modified by conscious modeling. This modeling effect can have several characteristics. In a novel situation, the effect will be somewhat delayed by the inevitable time needed before a new model can be formed. This is the situation in a young child, or an MLT with a limited memory system. If the interaction is similar to a previous one, a previously created model can be accessed more quickly from memory in forming the response. If the earlier model is based on frequent previous interactions we speak of a habitual response. This form of response becomes more and more common as we enter and live through adulthood. It is the basis for action based on "motor memory", "thoughtless habit" and prejudice in its broadest form of meaning – like our "taste" for certain forms of art, music, and smells and textures of food.

4b * MLT→NLT

This relationship is not substantially different from the one seen in RLT→NLT, except that the applied stimulus can be intentional. Intentional means that the interaction is initiated by the MLT for a purpose, which is an imagined outcome. (One should not forget, however, that the previous types of interactions can also occur when an MLT is involved). This goal directed behavior (action) by the MLT changes the probabilities of the potential outcomes. The total number of possible interactions between the MLT and the NLT has not changed, but the probability distribution of actual outcomes is affected. If the actual outcome of the interaction clearly approximates the imagined outcome, then we can say the model of the interaction created by the MLT was an accurate one. This kind of interaction is involved in human activities from chipping a rock to form a tool to creating an environment in a test tube that causes the creation of a specific set of chemical events.

5a * MLT→RLT

This interaction pair is fraught with great uncertainty. It is in some ways opaquer to our understanding than the RLT→MLT or even the MLT→MLT relationships. The reason for this is that we are more prone to imagine it as an MLT↔MLT type interaction than is warranted by reality, a reality which we can only ascertain very indirectly. The most important area in philosophy that deals with this conundrum is probably the field of environmental (non-human) ethical considerations. The closer a RLT comes to an MLT in behavior, the more closely we tend to be emotionally and rationally empathetic. Our inter-species versus the intra-species relationships, although in all other regards very similar to RLT↔RLT, differ from the latter by the inclusion of intent on the part of the MLT and the lack of the modeling capacity by the RLT needed for more than an emotional or instinctive reaction.

5b* RLT→MLT

Although we have some of the same difficulties of interpretation here, as in the MLT→RLT relationship, we are less concerned because we do understand that an RLT has no choice in its initiation of action. Our brain-mind activity will usually be more centered on the effect of the interaction on our being, rather than the imagined goals of the RLT. That is the main difference. In the MLT→RLT interaction the goal is perceived as central to the interaction, in the RLT→MLT interaction the effect becomes more central.

6 * MLT↔MLT

The egocentric goals and effects of each being share basic properties, but their actions alternate until either one or the other dominates, or a harmonious equilibrium is found. A search for an equitable outcome, especially in human-human interaction, is the ultimate ethical goal of wisdom driven action. It is the modeling of concepts such as justice in terms of their ontological, epistemological and relational underpinnings that is one of the tasks of philosophy. The highest wisdom in the mind of any one being is however immaterial. If that being cannot communicate those concepts to other MLTs and if they singly or together do not have the power to implement actions toward wisdom-directed goals, then wisdom comes to naught.

For wisdom to change the future it requires the use of power. The basis of the next two chapters is the need for accurate MLT↔ MLT communication, and the persuasive power it provides for the success of such organisms.

Chapter VIII
Wisdom and Its Needs

8.1 Communication and Wisdom

Communication is a form of interaction, specifically an action used to provoke a response. In this it differs from actions that are not intended for such a provocation, in which case it is only potential information. Usually the transfer of data, information and concepts is offered in an effort to change or stimulate the contents of the brain-mind of other(s). It is an effort to change what is known or believed. Communication has purpose, and is distinguished from other types of stimuli that are not driven or influenced by a goal. Also, communication presupposes an audience at which it is directed, including those in the future and those not physically present. There are therefore many levels of communication. The most intimate type is the intra-brain-mind kind, called thought, a recursive process. The next level of complexity is an interchange between a few individuals and is called a discussion. When many other minds are reached, without expectation of individual retort, we can speak of education or propaganda. Depending on intent the two overlap considerably. Philosophy uses all three forms of communication, and all, together, can be a source of wisdom.

8.2 Words and Their Concepts

Although discussed in passing in the part on ontology and again in the previous chapters on epistemology, words and their meaning need a more detailed discussion. Human thought and speech are near the end of a long string of emergent possibilities and they are central to any understanding and discussion of the human condition.

In the earlier discussion of ontology of the brain-mind, and the emergent possibilities of that system of systems, I limited it to a review of the subsystems the major systems, the I-E and C-L (see § 12.1). They were: the abstraction of stimuli; the use of imagination in forming a model; the storage and retrieval of some of these models in various memory systems; and the short temporal basis of knowledge compared to information and data. We now continue by delving into communication at greater length. To recap: A word is a symbol for a concept and the meanings of the various parts of speech are based on the actual relationships in reality, as modeled by the brain-mind.

Most of our conscious thought is a flow of words and it is difficult for us to imagine thinking without them. However, words could not have come before the concepts they express. As noted in passing in the previous section, words are an extension of the communication armamentarium using sounds beyond grunts, screams, chatter and various soothing sounds. Initially they expressed, in more nuanced terms, the emotion and instinct driven warnings, attractions and calming information. This allowed for more specificity as to source of and type of environmental interest. For example, the warning sound for snake versus lion would be different. As our earliest human, or even proto-human, ancestors developed the neurological underpinnings for better long-term memory, as well as greater ability to differentiate between similar environmental conditions, the need for an increased variety of sounds became not only useful, but necessary. For this to happen, a parallel change in our vocal apparatus needed to occur. These parallel changes undoubtedly accumulated only over thousands of years by the process of evolution.

Modern speech is composed, in all languages, from a relatively small set of syllabic sounds which differ by small and large degrees between language families. The exact enunciation of the specific set of sounds is learned early in childhood, and fluency becomes more difficult in later years. Early exposure to a bilingual environment can make the later acquisition of new linguistic sounds more attainable. Here I am speaking of the pronunciation, not vocabulary of a new language.

One difficulty with a limited set of sounds is that it gives rise to the fact that the same sound can refer to several, often unrelated, concepts. If the spoken word has distinct but related meanings, it is said to be polysemous. If the meanings are unrelated, they are called homonyms.

Finally, and unfortunately, the underlying concept evoked by a sound may be unique to a person or group. All of the above, together and separately, then give rise to misunderstandings, both small and large. Misunderstandings lead to confusion and conflict rather than clarity and understanding. Any field of investigation that seeks to understand reality therefore needs a common vocabulary and set of non-verbal symbols if it is to progress.

There is yet one further major difficulty that is not as easy to overcome by the use of common symbols for the same concept. The problem is that a concept that is common in one linguistic and cultural group often has no equivalent in another culture and language. The examples are numerous. Some are relatively benign, as the lack of the same concept, which in German, is called "Gemütlichkeit", does not exist in English. On the other hand, the underlying concept that in English we call "human rights", indicating personal rights above the rights of the community, is not part of the concepts held in China. This has led to many disputations as to the "correct" interpretation of the United Nations Human Rights Committee's works such as the International Covenant on Civil and Political Rights. This is one of the reasons why ethics vary from culture to culture, as will be discussed in Part III of this book. One hopes that philosophers from various cultures can come to a better understanding of these difficulties than politicians, lawyers and theologians — most of whom proceed on the supposition that they have the "correct" view. But this requires a search for better methods of communication, rather than disputation.

8.3 What Is Meaning in Communication?

In the sentence, "What does 'X' mean?", X can be a word, which is, as we have seen, a symbol for a concept. Mostly, there is not one meaning, that is, "the meaning", of any concept, word or action. When we try to enunciate what X means, we usually rely on describing consequences of the actions on the world— potential or imagined – including any changes to a brain. The consequences on the brain-mind are often not acknowledged, if even thought of. I propose that the meanings of X are all the consequences that will occur in reality, whenever it refers to a specific word, phrase or argument. The same is true of all other interactions at all levels of complexity, but communication is at issue in this section. Our understanding of the meanings of a word is based on previous results, as well as the imaginative foretelling of possible future effects. This interpretation is supported by the fact that if the word, "ibnaxoq", does not lead to at least one specific expected consequence, we call it a nonsense word. Of course, the use of any nonsense word also has consequences, but they are not specific to that symbol. The same is not true if the word is known to be from another language, in which case the result could be puzzlement and as well as a lack of communication. The intended meaning would in this case apply only to the speaker.

Sciences, technologies, and special areas of academic interest have dealt with this problem, as discussed previously, by creating a common vocabulary. This vocabulary changes over time as new information and understanding create the need for adjustments. As was also noted previously, the same intensity of effort to make such adjustments has not been forthcoming in philosophy. A major reason for this lack of conformity is that there is no agreement on basic ontology and epistemology. This was forgivable in previous eras when information was solely based on personal experience, using only the typical senses available to mankind. If the purposes of philosophy do not include attempts to clarify what words, concepts or relationships mean, then what are we left with? We are left with confusion, animosity, and hubris — not wisdom.

If those that labor in the field of philosophy cannot, together, clarify the basic, most common meanings, as defined above, of such words as justice, free will, purpose and beauty, then no progress is possible. To use a metaphor, it is as if many farmers, tilling the same field, are planting many different seeds, without order, without regard to what others have done, ripping up what they do not know-and calling it weeds, and are not getting together in order to grow useful or aesthetic plants. In the worst cases, some of these "farmers" may insist that a pebble is a seed, that acorns will grow into roses or that the meaning of being a farmer is only to work the field by turning the soil.

In building such a system of words and meanings, each case must be pursued by recursive methods of building from the simple to the complex and then dissecting the complex into more simple components, and then do it again and again as new insights, information and consiliences are found. This differs from using the method of the dialectic, an apposition of opposites. This will not work in most ordinary cases because reality does not just contain two possibilities when describing anything more complex then perhaps some interactions in subatomic physics. Even in that case, probabilities, not certainties, are the best estimates that can be used for what the meanings of any X could be.

8.4 How to Deal with Changing Meanings of Words

When words are used in everyday, typical discourse, there is only a small set of meanings to which the majority of the speakers subscribe. But even here change is a constant activity. A few examples will suffice to make the point. Nouns that refer to common things in the environment are the most resistant to change. The referents of the words dog, cat, stone, finger, sky, cloud, water, and fire have been the same for thousands of years. There are also things that can be experienced directly by each person such as joy, boredom and pain. On the other hand, words such as molecule and planet are recurrently associated with new and improved concepts. They are then adopted into the common vocabulary from the linguistic world of science. Historically, we can also describe the changing meanings in closely related languages and dialects. The same root has led, over time, to related but different meanings. In the Anglo-Saxon language family, we have the word "stool", which in English refers to a seat without a back, whereas the German homophone, "Stuhl", refers to a chair — a seat with a back. Both however have the same meaning medically; in that they may refer to excrement. The German word for dog is "Hund", whose near homophone is "hound". Whereas the former refers to the broad category, the latter has come to mean a tracking or hunting dog, a narrower meaning. The examples are legion, and in English, we also have the inclusion of many words from the French, from the time of William the Conqueror. Furthermore, all three languages have many roots in Latin or Greek, but with divergence into different meanings. Latin and Greek are also favored sources for neologisms although the new meanings are not necessarily always consistent with the original usage.

Adjectives and verbs are more subject to change and to an increase in the number of meanings they can have. Those meanings are often clear in context, and can be purposefully used for humor when the context is not given or is misdirected. The word "blue" refers not just to a color, but also to a mood. In German, the similar color word for blue, "blau", can also refer to a drunken state. The word "gay", once referred to a carefree and happy state of mind, but later became a descriptor of sexual preference. It is small wonder that translation is such a difficult process. It also underscores the need for words, that when used in the technical sense, are not tied only to a particular language. By using technically defined words translation becomes less fraught, and the result is less likely to be erroneous than if translated into phrases using everyday language.

Wisdom is only of use when the typical human can understand its intent, its source and underpinnings. There is no wisdom in misunderstanding.

8.5 The Purpose of Wisdom

The study of purpose is called teleology in philosophy, is another emergent property of mind. The concept involves an imaginary and desired future state of affairs, guiding the actions of a model forming being.

An amoeba, a one celled organism, has no possibility of creating concepts and symbols. But as we humans contemplate such a life form, we have great difficulty in avoiding assigning such concepts as purpose to it. Commonly, this is caused by misuse of the verb "to want". If we are to understand, and agree upon, what the concept of teleology implies, and what kind of thing can create it, we must tackle its meanings.

Purpose refers to a broader concept than the word goal does. The latter typically refers to a set of clearly enunciated complex relationships imagined at some future point in time and space. Target, is an even more concrete term in its referent, implying greater specificity. They are, and can be casually used interchangeably, because the indicated future state of affairs is usually defined. Purpose is a concept that underlies and is central to many philosophical debates. Free will, morals, ethics, aesthetics, and justice – all make use of references to consequences, that is, the future state of affairs engendered by past activity. Purpose cuts across all these topics and is also discussed in §§ 11.5 and 13.1.

The consequences of energy transfer can easily be described using terms built up from our understanding of the physical-chemical underpinnings of our existence. The formation of a molecule from its constituent atoms; the formation of order in crystals; the disruption of order by applied energy leads to a shattered crystal, to earthquakes, or to a supernova. All can be understood as purposeless despite being possible in a universe without humanity or any other model forming living thing (MLT) (See § 7.7 for review). In the process of trying to understand and describe the actions of RLTs and their relationships, the question becomes whether an RLT is complex enough to invoke purpose and all the other concepts that contain purpose as a constituent. In a non-transcendental worldview, the answer can be shown to be "no".

If that is so, then an MLT is necessary for the concept of purpose to make sense. That furthermore implies that these concepts do not refer to anything that is not an MLT. Also, since each MLT has at least a slightly different set of concepts, the best one can do is to discern the necessary and sufficient assembling commonalities. Only then can this concept have any bounded meaning so that we can continue on toward an understanding of the more complex terms and phrases. If one should choose to use some definition not based on such a process, one ends up with a muddle of one's own devising.

What then are those commonalities underlying a useful and precise definition of the word, purpose?

"Purpose: an imaginary and desired future state of affairs which guides the actions of a model forming being."

This definition involves all of the previously discussed aspects of the ontology of this universe and the epistemology of an MLT, most especially a member of our species. The necessary and sufficient mind processes needed to create a purpose are:

1* Abstraction — the conversion of stimuli to new matter/energy forms based on pathways that conserve the salient features of the source of the original stimuli.

2* Imagination – the (re)arrangement of the products of abstraction into new configurations with the addition of input from memory, both short-term and long-term.

3* Emotion —reflex reactions to the products of the abstractions and the imagination at the sub-and pre-conscious level of the nervous system. (Emotion, as a major factor in "free will" will be discussed in Part III.)

4* Time, space and the rules of cause-effect relationships emergent at each level of complexity.

The final product, a purpose, comes from an assembly of all these factors by the process that creates models, central to which is imagination. Imagination is an automatic preconscious process with possible conscious feedback that compares the imaginary outcomes of various models of action. The relative importance of each mind process will vary due to many factors. What these factors are, and how they affect the outcome of the evaluative functions of the mind will be discussed in more detail in the next part of this book.

The evaluative functions, especially those with a large input from the emotional brain, have been the most difficult to deal with when attempting to provide a consistent worldview. The reasons for this are similar in all cases. The first is that the separate reactions of the emotional and cognitive brain-mind are created by different pathways and have served different functions in the evolution of RLTs, and eventually the surely most conflicted MLTs, the human species. The emotional brain reflexively responds more quickly and less specifically to the environment. The opposite is true of the cognitive mind that utilizes imagination to compare possible outcomes of envisaged responses. In the cases, when the short-term utility conflicts with a long-term utility, and the long-term utilities are unclear, then indecision, stress, impulsiveness or a fallback to the reactions based on the strongest emotion may results. Since many people may respond to the same set of circumstances, each with a broad variety of possible outcomes, then one of the questions is "How is it possible that we agree on anything?" Some agreement can be found in the evaluation of negative consequences as well as in positive evaluative responses. One answer, from both a Platonic and theological point of view, is that there is another level of reality that determines what is good or bad, beautiful or ugly, just or unjust, etc. and that the answer is fixed by that level. The answer from a more Aristotelian, scientific realism or hypognostic worldview is more complex, nuanced and less certain. In short, the answer comes down to the fact, that the genetic plan that underlies the structure of a typical brain of any one species is at least 99% the same, especially in the configurations that govern both reflexive and preconscious activities. Conscious activity is much more highly affected by the environment, leading to different memories and different habits. Nevertheless, there is a large component of commonality that can explain common responses to common environmental conditions, especially those that have the potential for large negative or positive impacts on the individual and the community.

This is why we agree, to a large degree, to what these evaluative concepts mean. For a short, but complete discussion of the specifics of the genetics of brain development see on internet: "Brain Basics: Genes at Work in the Brain": http://www.ninds.nih.gov/disorders/brain_basics/genes_at_work.htm.

A decision is called wise if a group of affected or interested persons agrees that the proposal leads to the "best" outcomes with the "best" methods of achieving them. This result depends in the judgement of each individual. How these judgements arise is discussed in Part III of this book: "Evaluations".

8.6 Creating Purpose in A Hypognostic State

Once comparisons of the models of possible futures have been made, they can be compared to our purposes. The first step is finding a "best available" model that satisfies both I-E and C-L scrutiny. If the comparisons of the models created by the two mental systems does not lead to one consilient model, one can actively seek more information, or put the problem of finding "the best" model aside. In the latter case an adequate, but acknowledged improvable model, can be a guide for our next actions. This current best model can be held in a memory (mental or electronic/written) for future review. One may also lay aside the search for progress in concept formation in order to implement currently needed or unavoidable actions toward old aims, even if these actions cannot stand up to full scrutiny. If no immediate action is needed, yet taken toward unquestioned aims, then this last state of affairs may be supported by active wignorance, the willful aversion to information. In all cases conscious actions toward a goal are attempts to effect changes in the world and are given force by a process called 'will'. This concept is mainly explored in discussions of what the term 'free will' refers to and is important enough to warrant a separate discussion. (§§ 13.11,13.12)

.PART III
Evaluations

Chapter IX
Beyond Reflexes and Instincts

9.1 The Emergence and Influence of Thought
The most notable difference of the modern human brain from those of other modern primate species, and especially all other animals, is the large size of the prefrontal cortex (PFC). This is the area behind the forehead, above the eyes, and is the site of the so-called "executive functions". These abilities are necessary for switching between different tasks and learning. Another well-known function is the ability to focus on one of several competing stimuli, an ability which is weak in attention deficit disorder. (Other areas of the brain are also affected in this disorder). The precise locations of this and the other brain systems are important for comparative and evolutionary neurology — and brain surgeons, but not so much for the description of their functions which are of more interest to psychology or philosophy. This anatomic information is, however, very important for the understanding that many systems exist and interact. The answers to the "which" and "how" continue to be under active neuro-scientific investigation, with much current progress. The deep questions of how to define thought, knowledge, emotion, etc. cannot be answered on the anatomic level of organization, which only covers where activity takes place during these mental functions. What these functions are must be sought by studying their range, limits and interactions. What the fine-grained actions are,

biophysically and biochemically, are the hardest questions to answer at this time.

Moreover, on the species level we can begin to determine how the various abilities relate and interact. We have learned that part of the function of the PFC includes the control of voluntary actions, working memory (includes short term memory and processing of that information), and cognitive flexibility (the ability to switch between and compare different thoughts and means). These terms used by cognitive scientists are about what the brain-mind does and what the subsystems contribute. The how of these functions is, at the level of neural networks, much more complex and interactive than the flow of electrons on computer chips. Unless you are a computer hardware engineer, you probably have a low level of appreciation of that relatively simple process. The greater magnitude of the complexity of the brain-mind explains the longer timeline for advances in understanding. As noted earlier, our brain-mind works on principles of both binary and analog systems of energy flow, and the interactions at the neuronal level consist of billions of parallel processing steps.

The cortical areas do not work alone, but are additional systems that enhance and alter the abilities that emerged previously in the evolution of the midbrain. Anatomically, the prefrontal cortex (PFC) is not well differentiated from the adjacent parts of the cortex by a deep sulcus (think mini-canyon). It abuts the other main part of the frontal lobe, the premotor cortex and primary motor cortex, which control skeletal muscle movement. The abilities that emerged due to the evolutionary growth and development of the PFC allow us to manipulate our mental models and thereby our interactions with the environment, including with others of our species. It is this conscious manipulation and evaluation of the mental models that we call "thinking" as opposed to mere perception and consequent reactions.

The evolutionary development of the PFC possibilities, beyond the abilities of our pre-H. sapiens forebears, spanned millions of years. The earliest use of improved stone tool production and use by early hominids goes back to at least 2.6 million years ago (mya). Starting only about 8.3K BCE this so-called Stone Age was finally succeeded by the Bronze Age when the smelting of metals was developed. However, there are many other documented improvements in tools, broadly defined, during that long intervening time. Although the archaeological remains of these are much less durable, there is early evidence of the use and control of fire (~ 1.6 mya), the use of bone and antlers for tools (~1.5 mya), the building of shelters (~0.5 mya) from naturally available components, and the development of pottery and weaving (~ 38-24K BCE). The making of body coverings from animal hides was also necessarily developed, according to climate and circumstance, over these years. Obviously, this progression of skills is the subject of broad scientific archeological research, and the many details, including time lines, can only be superficially pointed at in this book.

It is, however, clear that these changes in mental abilities, which led to increased learning, invention and then communication of skills, did not occur all at once. But because many of the archeological remains that point to these developments were less durable, there is less hard (no pun intended) evidence of the exact time line. In the case of transient interactions that left no mark, such as the development of language, we need other sources of information and the use of inference and deduction. But the current paucity of information from archaeology and comparative neurology is no reason to invoke a mysterious force that separates mankind from other life forms.

These most useful sources of further information are of very recent origin. They include: interspecies DNA analysis; comparison of nervous system biochemical processes; the comparative embryology of nervous system development and abilities in our closest nonhuman relatives; and finally, the study of cognitive development from infancy through adulthood in typical and atypical or brain injured human beings. When all this data is combined it paints a picture of a gradual, complex process of cellular differentiation, growth of neuronal systems, and changes in the interactions and connections between different areas of the brain. All these physical changes have led to the emergence of newer mental capabilities and uses. In comparison, the evolutionary ability to repurpose older structures can be demonstrated more easily with the bony structures of the chordate line. One example must suffice for our purposes: the development of the tiny bones of the middle ear from much larger precursors, originally part of the jaw. For detail see, for example, the Wikipedia article "Evolution of mammalian auditory ossicles".

Among the products of the interaction of the amygdala and PFC are the concepts of art, religion, ethics, justice, free will and other such concepts, none of which are present at birth. They only emerge as the necessary brain areas grow in size, the complexity of the interconnections increases, and the environmental inputs alter both.

In humans, when the pre-and postnatal nervous system developments are atypical, we note helpful and harmful changes in personal and social abilities and behaviors. On the positive side we observe the talents — referring to exceptionally well-developed abilities in such areas as inventiveness, the arts, mathematics, linguistic abilities and motor abilities and social skills. The more problematic changes include some mental properties that are not normally seen as talents, but are often attributed to be solely environmentally dependent and learned abilities such as empathy, levels of aggressiveness or pacificity, motor skills and others. However, observation of toddlers and young children show that underlying tendencies in these areas, that is talents, exist. But, just as typically recognized talents can be honed to higher levels of ability, left to languish, or be moderated by conscious effort, so too can the socially important behaviors and tendencies. This honing is one of the major contributions of the non-instinctual brain, especially the PFC. This is why the old arguments of nature versus nurture are at best misleading. Only an understanding of the combination of impacts from both nature and nurture will lead to more complete and accurate models. In any one person, at any one time, it is the relative correctness of the models, as developed in a brain-mind, including emotional states, that determine the extent of possible wisdom. Empathy, a form of emotional mimicry, is especially important pertinent to philosophy because it greatly broadens the sources of input, even if they are experientially one step removed. It is a mental response mirroring the perceived emotional and physical state of another organism. It is especially felt when the particular circumstances are also perceived directly. The reactions of pain and fear in another being are probably the strongest stimuli of this response. As an example, I personally react to the vision (or even thought) of a fishhook in someone else's finger with both an emotional jolt as well as a motor twitch of avoidance. This ability to empathize is very likely an evolutionarily useful trait that facilitates 'learning from the mistakes of others' without necessarily being in personal danger. An extensive discussion of the pertinent research and understanding of this topic is found in the recently published "Mama's Last Hug" by Franz De Waal (2019).

Thought then is the conscious awareness and evaluation of our internal and external environment. It is composed of many threads and pathways starting with stimuli perceived with our external sensors, and the input from the circuits involving memory, emotion and complex reflexes such as empathy.

9.2 Wisdom –A Product of Evaluation

We continue by reviewing the definition proposed in Chapter I. Wisdom: "A product of a mind, based on a highly informed understanding of the constituents and relationships of our world, which in turn leads to refined predictions of which possible actions lead to the highest probability of achieving specific goals.

Approaching the ideal end result by the evaluation of possible choices is the product of wisdom. It is ethically neutral when the means and goal are as well. The two main mental pathways concerned are the Instinctual-Emotive (I-E) and Conscious-Logical (C-L) (See § 12.1). In everyday adult human activity, the simple instincts play a relatively small part in reacting to the environment, being largely overlaid by the emotional and conscious aspects of the brain-mind. The combined use of the latter two major pathways usually gives the best chance of choosing wise means to a goal and achieving them. Depending on either system alone is not only incomplete, but can be very misleading. Furthermore, all actions toward any one goal will also affect the probability of reaching other desired ends in an expanding cascade of side-effects. In the process of evaluating these secondary effects, some of the potentially affected results may be judged to be relatively inconsequential. Or, the imagined impact on other desired results may act as further input in judging the proposed actions so that a balance between them is achieved. The first conclusion then, concerning wisdom, is that it concerns a process which recognizes that all actions have broad consequences and that the search for wise actions implies an effort to consider the foreseeable results.

9.3 Why "Evaluations" Rather Than "Metaphysics" Or "Axiology"

Metaphysics is often the name given to a third section in a general book of philosophy. Historically, it was initially used by a "first century, C.E., editor who assembled the treatise we know as Aristotle's Metaphysics out of various smaller sections of Aristotle's works." (SEP, 2021). It referred to those subjects that were discussed after his "physics", or natural world, sections. Later, other authors tended to include the various subjects that did not clearly fit under the headings of Ontology or Epistemology. Usually, Metaphysics also includes the discussion of concepts derived from axioms based on the proposed reality of transcendental realms. This subset involves, among other things, the world view of religions. Such axioms concerning non-physical but immanent forces are excluded in this book, as they are not based on an ontology consilient with the confirmable data of science nor can they be investigated further.

One of the common definitions of Axiology is that branch of philosophy that considers the study of principles and values. However, in many cases discussions in this area do not stress that these principles and values are both biologically based. The biologic basis can be divided into the biophysical needs of any living organism and those needs arising to maintain harmonious mental functions. Evaluations encompasses both.

In the endeavor to find a common denominator among the diverse "after physics" topics, I found that the mental activities that they seem to have in common could be summarized by the word "evaluations". And furthermore, that these values need to be separated, for clarity's sake, into biologically pertinent values and psychosocial values.

A discussion of my choice of the word "evaluations", and related terms, is now in order. One of the best one sentence definitions I have found is taken from the current (2018) Wikipedia article. It is: "Evaluation is a systematic determination of a subject's merit, worth and significance, using criteria governed by a set of standards." This could be taken as referring to only conscious processes, but it can also include many of the unconscious processes as well. Most importantly for this discussion are the two major evaluation systems, the I-E and C-L. Both of these use inputs from the classical external senses, memory, instinct and habit. As used in this book the word "value" encompasses the word "worth". "Merit" is limited to a subsequent conscious evaluation of the source of input and the logical connections between these data and the resulting model concerning an imagined future reality. "Significance" involves the same imagined consequences, involving consciously considered connections between past, current and proposed actions. These connections are also the source of meaning.

The word "meaning" is often divided into three sub definitions in English dictionaries. Again, which one, and how, to use this word in philosophy is not clear from these sources. All of them, however, involve the idea of consequence. It is least often used (in philosophy) in terms of purely physical consequences, and then only in reference to living things. For example: what is the meaning of a forest fire? Most answers will describe the direct effect: destructions of trees, animals and the environment in general. However, it also means that there will be marked erosion with the next major rainstorm.

The answer to what is the meaning of the word, is found in its impact, as a stimulus, to a brain mind, whether human or trained animal. At best, a nonsense word is ignored. At worst, it causes confusion or misunderstanding leading to war. All foreign words are in essence nonsense words to a recipient ignorant of the users meaning.

Finally, the much-debated question, "what is the meaning of life?" This can be actually subdivided into two questions. The first is: "Why is there life and how did it come to be?". The answer to this is either a transcendental one, or a secular, scientific one. Either is at base an ontological answer, that answers the question "what were the necessary and sufficient conditions that had the consequence of life coming into existence". The second is: "What purpose, that is consequence, will (or do I propose) my life, or that of another being, have?" The final answers will of course be conditional until the end of that individual life. The import of this question will be given more consideration in the Epilogue, "A rational hope".

This brief discussion supports the idea that conscious evaluations are almost always recursive. Emotional evaluations are less so. In the latter the input the emotional brain-mind is based on instinctual reaction and (often) un- and pre-conscious memories. These then trigger a complex, reflexive output. The resulting emotional states are based on evaluations that are mostly hardwired, evolutionarily derived NS configurations. Merit is not a consideration in this case. The inferred significances follow from a conscious seeking for consilience with the recalled emotions associated with similar past experiences. The standards of a C-L evaluation are in turn based on the perceived significance of biologic facts or the merits of received cultural or personally developed norms. For emotional evaluations the typical responses are built into the brain circuits during the fetal anatomic development of the brain and the subsequent modifications caused by experience.

Mental evaluations are the necessary precursors of complex reflex actions, categorizations, as well as willed (conscious) choices. As with most activities of the mind, there are many circuits, small and large, and centers that serve a particular function. In this case the two major systems are the conscious and the unconscious processes. These two systems both take in stimuli concurrently, process them separately, and then, in subsequent, more complex operations, interact by furnishing input to each other in a recursive way. Roughly speaking, the conscious evaluations take place in the cerebral cortex, and the unconscious take place in the midbrain and hindbrain. For an extensive, very readable discussion of brain-mind activity, and its anatomic, structural and electrochemical underpinnings see Sapolsky's "Behave" (2018). A more neuro-physical discussion can be found in the book by Paul Thagard, "Brain-Mind" (2019).

In this book, examples of the two major forms of evaluation will be discussed. The first set includes the evaluation of aesthetics, exemplified by beauty, a mostly preconscious activity. The second is exemplified by planning, a process based on the C-L system. This can be illustrated by the following effort: calculating the shortest route for a series of errands. This is primarily handled by activating the imagination in the cerebral cortex, with the informational stimuli (pertinent information) retrieved from memory.

More complex is judging the ethics of a situation, which is an interactive process that uses input from those brain areas that generate instincts and emotions, as well as those that generate the conscious models concerning the possible outcomes of that particular situation. An ethical judgment can be wise in a C-L sense, but is more often heavily influenced by the I-E subsystems with minimal cognitive oversight. This is especially true of "snap judgements".

A short review of the brain systems involved follow, to refresh the memory of, and expand on, some of the main points concerning these systems as covered in Parts I & II of this book.

9.4 Ontology of Mind – Continued

The necessary cooperative efforts, between seemingly divergent fields of study, to progress in our understanding of mind is well stated in the following quote. "Although the ultimate aim of neuroscientific enquiry is to gain an understanding of the brain and how its workings relate to the mind, the majority of current efforts are largely focused on small questions using increasingly detailed data. However, it might be possible to successfully address the larger question of mind–brain mechanisms if the cumulative findings from these neuroscientific studies are coupled with complementary approaches from physics and philosophy. The brain, we argue, can be understood as a complex system or network, in which mental states emerge from the interaction between multiple physical and functional levels. Achieving further conceptual progress will crucially depend on broad-scale discussions regarding the properties of cognition and the tools that are currently available or must be developed in order to study mind–brain mechanisms." (From:" Understanding complexity in the human brain" (Bassett & Gazzaniga, 2011).

The chapters of Part I showed how we can move from atoms to the human brain-mind as the ever more complex nervous system allowed the emergence of new properties. A few definitions are restated here for the convenience of the reader.

Brain: an organ composed of nerves and their supportive cells in a highly structured, highly interconnected system evolved in organisms that survived in an inanimate, unmindful and indifferent environment. In pre-human primates this was followed by further evolutionary developments as our distant ancestors had to deal with the more complex behavior of other self-aware beings.

Brain-Mind: the sum of all brain activity beyond simple spinal motor reflexes includes, as follows;

1* An awareness of motor reflexes as the N.S. receives feedback arising out of the motor responses

2* The I-E (the complex reflexes) that are major informants to other systems, including the C-L.

3* Habits as formed by the preconscious activities of abstraction, imagination (reconstruction) and memorization. These types of functions are all reflex-like, but have both external and internal sources of stimuli that are generally processed in series. Once formed and laid down in memory, habits function much like instincts even though they are acquired by additional experience rather than being mostly genetic in origin. Furthermore, instincts can be overridden by habits and cognitive inputs over time.

4* Reactive consciousness; a more recent emergent property of nervous system evolution. It can be conceived of as a perceptual playback loop in Short Term Memory (STM). Though the mechanisms are not yet fully elucidated, functionally it is the brain state that allows: (1) the comparison of concepts and models, (2) their (re)storage in memory as well as (3) access to previously stored memories for response formation that is not sourced from instinct or habit alone, and (4) an awareness, but not evaluation, of our feelings.

5* in a typical human brain, the interactions of the I-E and C-L functions are so interwoven that it makes the study of the underlying sub-functions much more difficult. These interactions have allowed emergent processes beyond the capabilities of either alone.

6* Will. The driving force of C-L action, as discussed in §§ 13.11, 13.12.

7* Self-consciousness; the awareness and evaluation of one's internal feelings concerning the effects we have had, or may have, on the environment or ourselves. The process occurs in STM differing only in content from reactive consciousness. (See next Section.)

9.5 A Newer Level of Emergence — Self-Awareness

The steps from conscious to self-conscious life was created by the emergent properties that were as transforming as those from unicellular life forms to those with specialized organs. Although we can study the physical organization of current human brains, especially as compared to other mammals up to and including the other primates, the transitional forms are not available for comparison. Nor are their behaviors.

This aspect of the current typical human mind is understandably the most complex. For millennia humans have hypothesized what causes their thoughts and asked "How have such concepts as ethics, free will and beauty arisen"? Now we understand the physical brain much better than a century, or even a decade ago. This has meant that many of the old hypotheses needed to be put aside or adjusted. The new hypotheses required that they be formed by encompassing the understandings recently gleaned from all the sciences – from the most basic, such as physics and chemistry, through those from biology (biochemistry, physiology, evolution science, ecology), and to the newly burgeoning fields of the brain-mind sciences (neuroanatomy, neurophysiology, psychology, and sociology), to name but the most obvious. However, it does not need to be overwhelming. As first noted in § 1.5, what is most important for philosophy is not a memorization of all the details, but rather an awareness and understanding of the emergent possibilities at each level of organization.

The metabolic and physiologic functions of complex life forms evolved extremely slowly, over eons of time, from unicellular organisms to those forms that are comprised of specialized organs. The human brain-mind also did not evolve in only a few steps from the early primate forms to the capabilities we have today. Among individual typical human adults, the capacities for and use of self-awareness are not, as yet, as evenly distributed as the functions of, for example, the visual system. Self-awareness means that we are aware of our own agency. The levels of this type of awareness can be regarded as closer to the distribution of the talents for mathematics, music, language acquisition, fine motor control, etc. than to our basic sensory abilities. It therefore behooves us to approach human self-awareness with less surety regarding its level in the hierarchy of mind-brain functions, and its availability in decision making, than older philosophical and psychological hypotheses have. This will be an important consideration in discussing "free will" (§§ 13.11, 13.12).

9.6 Is Life Sufficient for there to be Evaluative Functions?
As discussed in ontology, life is a set of attributes of certain complex structures that include, at a minimum, growth, repair and creation of more complexes with the same attributes. This ability to re-create itself decreases the risk of extinction of any particular complex (organism) by spreading risks and opportunities among the many copies.

The only form of duplication relies on cell division. Later, in the process of evolution, cell differentiation allowed even more flexibility in the face of these risks and opportunities, and this emergent property eventually led to further variations on the basic patterns. Sexual reproduction promoted this flexibility.

Recall from § 7.7 the two types of life forms that inform the coming discussions. Humans are Model Forming Living Things (MLTs) and can have cognitive evaluative functions and purpose. The other type of living form, the Reactive (RLTs) can only have propensities. Therefore, the short answer to the question in the heading is a "no". Life is also an attribute of RLTs, but it alone does not confer the ability to create purpose. This is because the RLTs do not have the full set of the evaluative functions found in an MLT. To review, to be an MLT, the following, emergent properties are necessary, and then only sufficient in their combination.

These properties are listed here for review, as described in the section on Epistemology:

    1* Abstraction — the structured patterns of neuronal activity initiated by responding to sensory information derived from the real world, including other neuronal circuits.

    2* Imagination — the forming of new models based on the reconfiguration of the basic abstractions and memory.

    3* Memory — the repository of patterns and concepts derived from both abstraction and imagination.

4* Consciousness — the real-time mental activity necessary to create and retrieve detailed memories and also to create the concept of a future. These necessary functions do not allow us to speak of conscious evaluations in RLTs.

It is clear that, for RLTs, abstraction can occur without memory or imagination. The simple reflexes are based on nothing more. More complex reflexes and instincts result from the further processing of these abstractions by the complex architecture and actions of the brain-mind system involved. Their habits need some memory, but it is memory that is only accessed subconsciously.

9.7 The Emergence of the Evaluative Ability of Physiological Needs and their Fulfillment

As with the discussion of previous concepts, the evaluative ability of a brain-mind needs to start with a working definition. What processes and inputs are necessary? As noted above, conscious " evaluation is a systematic determination of a subject's merit, worth and significance, using criteria governed by a set of standards." (Wikipedia 2018).

This definition is limiting, in that it infers a self-conscious brain-mind when it speaks of merit and worth. Significance is a much more basic attribute, because it can rely on just the evaluations based on the system standards built into their I-E system. But, the evaluative determination of merit and worth depend on a standard created by reference to personal experience, as laid down in memory, as well as information socially transmitted by the passing on of these standards by words or deeds. The four concepts together are major inputs in estimations of conscious value evaluations. Such estimations are more fully discussed next.

9.8 Axiology and the Determination of Value

Axiology is the philosophical study of value, the latter having no standard definition. The word "scientific" by Robert S. Hartman (2011) concerns whether thing "α" is a good or bad example of a concept depending on how many of all definition criteria (properties) are met. How does this apply to actions? It seems that actions can be evaluated in two ways. Let us examine the action "raising my left hand". By Hartman's criteria, an action "α" is a good example if it consists of exemplifying the definitions of "left, hand, raising, my". This means it has to meet the criteria of the definition of the main words – i.e. left is by definition one side of a two-dimensional representation of a bisymmetrical object, seen from the "forward looking" perspective of the object. This definition in turn requires the description of its components, which then also have to be met if "α" is to be a good example. These regressive definitions can go on almost ad infinitum, but in common language terms they are easily understood and confirmable by ordinary observation. But in the discussions of such an action, as seen in a mirror, shows that confusions can arise especially if the image is not distinguished from the object. Furthermore, does the meaning of the whole phrase "α" include or exclude the concomitant raising of the attached arm? And so on!!

For ethics and aesthetics, the situation becomes even more complicated when using this method of determining whether "α" is "good" or "beautiful". This is because for an ethical evaluation the intentionality, and/or the meaning of the action on others, need to be considered as well. In aesthetics the meaning of the perceived sensory input, the effect on the observer, is of primary importance and it is primarily evaluated by the I-E system. In both cases the definitions require an understanding of the neurologic activity which can be said to underlie intentionality and personal meaning. (Chapter X "Value" & sec. 13.3 for 'Good vs Evil').

9.9 Defining Intentionality and Subjective Meaning: Some Models of Neurologic Activity

All conscious mental evaluations imply comparison with an available mental model to serve as a standard. Such comparisons can be made with recent similar sensory inputs in short term memory, as well as with inputs from long term memory of more temporally distant, learned standards or encounters. At a minimum this involves those parts of the brain-mind that comprise the abstractive sensory system, the creative imagination system and those that lay down memories and retrieve them.

Of these, the subsystems involved with abstraction are the best studied by scientific methods. The pathways involved in creating new models by the process of imagination and those used to transfer the resulting models to short and/or long-term memory have been much harder to tease out. But the most recurrent aspect of increasingly more complex neuronal systems is that they are first, built on earlier systems, secondly, interact with these progenitors, and thirdly, have a functional similarity between the older and the newer. Stacked systems are the rule rather than the exception. By this description I mean to evoke in the reader a recollection of their understanding of the functionally "cooperative" interactions of organ systems in an individual being and even more so the interactions and adaptations of an ecologic system. These non-neuronal systems are among the more commonly known complex natural organizations of similar intricacy. The amount of information needed to grasp the complexities of all these systems can be daunting, but attempts to advance understanding continue to be made.

After having considered the subsystems that underlie abstraction, imagination, the storage and retrieval of memory, as well as their interactions with the emotional-instinctive systems consilient, rational models of concepts such as beauty, ethics and evil can be created. But, none of the evaluative comparisons pertaining to the latter concepts can be made by quantitative measurement. Only by a ranking, within the group of each concept type, can comparisons be made. Such ordering allows for the creation of a relatively simple hierarchy based on only a few terms. This hierarchy is informed by qualitative variations in the terms that underlie the definition of a word referring to an evaluative concept. The level of beauty "X" versus beauty "Y" can then be approximated by assigning relative values to each term in the set. Namely the content of a particular aspect may be: not in set, lowest, lower, equal, higher or highest. Symbolically this progression can be written as 0, <<, <, =, >, or >>. Furthermore, each comparison should only be made within each subcategory of the necessary qualities that compose the set of properties. All these necessary terms for each definition form the minimal set. The assigned values may incorporate more than a simple scale of: "0, << ..." but the lesser will suffice for this discussion. Also, the terms of the definitions must be made explicit as to their relationship to true synonyms (tautologies) versus nuanced similar terms. This is not a trivial task.

To organize a concept by including more than one property at a time, one needs to go beyond the one-dimensional evaluation that is the limit of a linear comparison; one must go further. Typical graphs compare two dimensions with an X and Y axis familiar from elementary algebra. In a cubic structure one can show relationships of three dimensions (properties), a feature of solid geometry. But, at that point the simple terms, such as lowest and highest, must be abandoned for the concept as a whole, and more use be made of an extended description. Beyond three dimensions such organization is likely to confound the typical, unaided mind in any attempt to model the concept without some help from mathematics or logic.

Multiple two-dimensional sets can be displayed and discussed by using overlaps called a Venn diagram. Per current Wikipedia entry, "A Venn diagram (also called primary diagram, set diagram or logic diagram) is a diagram that shows all possible logical relations between a finite collection of different sets." What organizes these sets of concepts is the exclusion and inclusion of specific qualities for each subset. Once the minimal necessary and sufficient qualities for a given set are named, and the main differentiating exclusions are noted, we can see the difficulties in evaluating complex terms and processes. We can rank and compare each quality of two actual sets of the same concept but find it much more difficult to inductively rank the actual sets as a whole. There are two exceptions to this. One is the ranking of the two sets, one of which has all the lowest and the other all the highest rankings. In the usual case, however, the rank of the pertinent qualities is mixed. However, there is a helpful next step. We can attempt to rank the importance of each quality, then give both the quality and the level of each quality an arbitrary number value (based on rank) and then perform a weighted average calculation.

A well-known example of this method is how final school-course grades can be determined and students ranked on total performance. For example: homework may (arbitrarily) count for 10%, quizzes for 20%, term paper for 20%, tests for 45% with 5% being reserved for subjective adjustments such as class participation. The percentages are then multiplied by the average value gained in each subcategory and then added together to give the weighted average. The weighted average for all the students then ranks them for the course. Of course, for many of the concepts of interest in philosophy both the percentage assigned to a quality and the absolute values may have relatively few possible assigned values- say from 1 to 5 rather than 0 to 100. This allows for more 'ties' between the compared specific actualities in question, without them being the same. But they are equivalent when one must choose between them. Therein lies the problem in making value choices between two similar instantiations of a concept. Even when based on the best attempts at deriving, for example, the relative values of the necessary physiological and social needs of all living beings, in a particular circumstance, it is a fraught effort. Hence the unavoidable subjectivity of non-metric evaluations.

Which is more important: food or water? Answer – "It depends!" The obvious variations in the possible states of hydration and/or caloric needs make an immutable hierarchy between the two impossible. Why would anyone expect something to be "set-in-stone" in the valuations of art or ethics; never mind the worth of individual human beings?

9.10 Thoughts on Evaluations Based on Psychological Needs

At best, such actual evaluations would be logically rational, although the assignment of values would be based on different subjective estimations. The perceived psychological needs and/or availability of resources (the properties of an instantiated or theoretical case) lead to valuations in these realms that are highly influenced by the structure, function and history of any one brain-mind. This leads to the conclusion that comparing these types of values is at best described as an ordinal and partially subjective endeavor. The comparisons must perforce be situational in the broadest sense. No absolute, objective, and cardinal, answers are possible.

Psychological needs, are often based only on the complex reflexes or emotions. But they can also be based on cognitively derived purposes, which by definition are derived from an imagined future. The C-L system also, if given time and attention, reviews the output of the evaluation of the I-E system.

Theologians assume and axiomatize the goals of a higher, usually transcendent, being. Mystics are similar to theologians, but usually have a less defined structure in their spoken and unspoken assumptions and goals. The self-aware fantasist, such as science fiction novelist, often makes do without any axioms whatsoever, or creates them (im- or explicitly) for the world faced by the characters. However, if fantasists, when facing reality, do not care to explore their implicit axioms, they are guilty of wignorance or self-delusion. A hypognostic realists, knowingly, is also ignorant of vast informational gaps. In this case the creation of their evaluations would be based on a search for, and consistent with, the current elucidation and applications of natural laws as derived from the study of the emergent properties at each level of organization of matter and energy.

For the hypognostics, the evaluations are a first step in the creation and understanding of possible purposes. For them, such purposes, and the means toward that end are desired to be consistent with natural laws. These laws are in turn based on formulations consisting of the inferred necessary and sufficient kinds of matter/energy as well as its organization and subsequent emergent properties. These formulations are expected to be subject to adjustments, using new information, as it becomes available.

The two types of fantasy-based evaluations (noted above), base their models of the world on unsubstantiable revelations and confabulations or on created axioms inapplicable to reality. Neither proposed scenario is supported by confirmable data nor can it be usefully compared or evaluated in order to arrive at supportable real-life conclusions. Although their derived purposes may be compared by the application of logic, the insupportable input gives rise to an unknowable truth value. Or, as inelegantly stated in a computer situation, "garbage in – garbage out". In logic the phrases would be "truth in – truth out" and "falsehood in – unreliable answer out". This occurs whenever the stated or implied axioms cannot be tested by the use of any agreed-to means. Fiction books do not pretend to discuss the truth. Wignorance also cannot be argued with, as it is based, at best, on self-selected partial knowledge, gathered to conform to personal, potentially irrational, means and purpose. If we are to agree that evaluations of philosophical concepts need to be based on, and be logically coherent with reality, then the approximations of rank should not run counter to the best measurements of the underlying emergent processes. These in turn need to be consistent with the rule-based ontology of matter, energy and emergent change.

If one argues that beauty, justice or evil are not based on physical reality then one should subscribe to the 7th Proposition made by Wittgenstein in his "Tractacus" — "Whereof one cannot speak, thereof one must be silent ".

However, following his proposition may not always lead to emotionally satisfying conclusions. It may even lead to despair if we look for absolute answers to the questions raised by these subjective concepts. In our search for answers, we must remember that, according to a secular, hypognostic world view, as well as the view of the agnostic and many theologians, we do not and cannot know the ultimate basis and future of our universe, nor our place in it. But, with this acknowledged limitation we will now discuss how the evaluation of the common psychological concepts that have bedeviled philosophy are processed by the human brain-mind.

9.11 What are the Kinds of Evaluations Performed by a Nervous System?
There are basically three kinds of evaluations, each based on the levels of the NS involved. As previously noted, the more complex levels almost always incorporate some input from those evolved earlier. A very thorough discussion of this topic can be found in Sapolsky's "Behave" (2018). I believe the discussion in this section is consistent with the extensive findings described

therein, and sufficient for the philosophical discussion at hand. The main divisions of labor in reacting to the world are:

1* Unconscious feedback loops. This non-cognitive, reactive process is seen in such basic physiologic processes as blood pressure control, management of digestion, heart rate and the more complex functions, such as some sexual behavior in animals. These functions are usually described as originating in the autonomic nervous system located mainly in the hind and mid-brain. These may overlap with the next system.

2* The preconscious responses that create effects that can simulate the conscious responses to the inputs from the senses. Reflexes, instincts, emotions and habits belong in this group. For each there is a minimal necessary input and the responses range from none to all, usually with a graded response to some maximum.

3* Conscious "thinking". This comes in two forms when applied to evaluations. Nonmetric that is rank (ordinal) order, or metric (cardinal) order. The former allows logical calculations and the latter allows mathematical calculations. Ordinal relationships are described in the written form with the use of logical notations as found in a book on formal logic. Logic consists of concepts (and their formal signs) such as less than (<), greater than (>), iff (if and only if) etc. For an extended discussion of how to use such notations in philosophy, the thoughts of M. Bunge (2003), Mahner (2001) are a good start.

External inputs are metric due to the quantum character of the matter/energy components. In the nervous system, however, the output on the neuronal level is binary, that is, the nerve either fires (propagates a signal), or not. The number of possible transmitter molecules released at the dendritic junctions is on a limited continuum and there are numerous individual neuronal receivers (dendrites) that receive input from the preceding axonal binary output. Each nerve also has a finite number of various, though specific, receptors (molecular structures), that respond to the transmitter molecules. This results in system outputs that are temporally and spatially variable, subjectively analog, although within a limited range of possibilities.

For example: In simple muscular reflexes the receiving muscle cell (myocyte) essentially reacts in an all or none manner to its stimulation by one efferent neuron. However, the muscle bundle as a whole gives an output from

Beyond Reflexes and Instincts 153

the minimal of one cell to a maximum of all cells in the muscle. Rarely are all the myocytes stimulated to contract at the same time, but with high input rates the overlap of action increases the strength and duration of the whole muscle's contraction. This is then subjectively perceived in an ordinal way, from barely noted to maximal exertion. As a result, any motor reflex reaction can be consciously noted on an ordinal scale, between no response and greatest response. But in fact, at any point of time there is some cardinal number of myocytes that are in a state of contraction.

At each level of complexity there are both relative advantages and disadvantages for the organism when using a particular subsystem in an interaction with the environment. The reflex evaluations, both simple and complex, were developed early by the processes of evolution. When cognition became possible, it allowed more specific and nuanced responses. In the outline below, by "positive" we mean that a response will usually help the organism to survive, grow and reproduce. The "negative" is, in a typical situation, more likely to interfere with those functions. This table is a simplification, but covers the most pertinent aspects of each kind of response.

(Table II starts on next page)

Table II – The Evaluative Systems
Instinctual-Emotive (I-E) and Conscious-Logical (C-L)

(+) Positive	(-) Negative
I-E/1* – Reflex	
Very Fast	All or none response
Usually, appropriate	No time for conscious control
I-E/2* – Motor Instinct	
Handles more complex input	Fixed type of response
Usually, appropriate	Cognitive control possible only *after* initiation of actions is noted by consciousness
Can be more nuanced because more information is abstracted; no mood or emotional component.	
I-E/3* – Habit	
Automated responses based on previous successes; often unmonitored by consciousness.	
(NOTE: Habit learned through conscious practice, thereby differing from Pavlovian (P) or Skinnerian (S) reflexes which are based more on repeated interaction of the environment with either the autonomic nervous system (P) or built on instinctive behaviour (S). Appropriateness of such behaviour is intermediate between instinct and conscious evaluation, but greater than reflex emotional or motor response alone because of previous conscious input).	
I-E/4* – Emotion	
Fast, broader range of intensity	Often inappropriate in type or duration of response
Coordinates responses by different systems	
More sensitive to cognitive control	
(NOTE: Emotions can be initiated by conscious input, although usually resulting in lesser intensity. The related concept, Mood, lasts for a longer duration than emotion, and usually of lesser intensity. There is no clear differentiating border between them that is not linguistic). For emotions and moods, there are multiple possible inputs (scenarios), and therefore more circuits and systems can be involved, than is the case in the first three systems. Because this reaction is initially of the reflexive type, there is no initial conscious evaluation concerning the source, the actual intensity, or possible contexts of the stimuli. These subsequent conscious evaluations take more time, and can imagine multiple possible outcomes for each of the imagined responses. Therefore:	

(+) Positive	(-) Negative
C-L/5* Conscious evaluation	
Recursive correction over time	Slow compared to 1* - 4*
More nuanced appropriateness of response is possible	
(NOTE: Belief based memory input can be a confounding factor depending on the source of the belief, and can result in either a helpful (appropriate) or harmful response.)	
C-L/6* Calculated evaluation (maths and logic)	
Best accuracy in reaching an appropriate response	Slowest
Data driven	
(NOTE: Awareness of their presence can make instinct, emotional and unsupported beliefs removable from these calculations. Data may however be given more weight than warranted, especially when it is not appropriately vetted, and can lead to the worst response when the data is false. Data may also be attributed to a non-existent cause-effect system and then may or may not lead to a correct evaluation. The abstractive processing of sense data (including machine gathered data) may also be faulty, and may lead to inappropriate imagination-based models, false memory and hence incorrect evaluations.)	

All the evaluative methods outlined in Table II may function concurrently and/or in series and therefore lead to variations in response. This is despite the same input of stimuli at different times, into the same organism's brain-mind or those of others. Objective data usually will lead to more accurate models. Objective are those aspects of reality that are measurable against an external (to the brain-mind) physical standard, one that is based on the properties of matter/energy at the various levels of organization. In those cases where data input consists of stimuli originating in the novel actions initiated by other brain-minds, these inputs may often be more accurately processed by the reflexive, instinctual (intuitive), and emotional systems. Our conscious system can subsequently only supply rational limits to these evaluations, based on previous consciously processed experience under (assumed) similar circumstances. These rational limits can delineate both minimum and maximum possibilities, as well as be used to judge the probabilities of the correctness of the more automatic evaluations. However, the absolutes of a rule-based mathematics cannot be applied and that always leaves some room for honest disagreement concerning the meaning of human action. Stopping at agreeing to disagree, however, can never lead to more accurate evaluations!

If the underlying axioms of the discussants are at irresolvable odds, or wignorance is present in any of the discussants, then a productive discussion becomes impossible. Since philosophy is a search and desire for wise decisions, based on the best concepts, then the ultimate goal involves greater and greater quantitative data acquisition, as well as diverse evaluations, that together can lead to more accurate predictions of the outcomes of proposed actions. This is the function of an education system based on an informed understanding of the world.

9.12 What Underlies the Evaluations of Human Interactions?
Conscious-Logical evaluations are arguably the mental functions that are the most typical and developed in humanity, when compared with all other organisms. All the other functions we share to a lesser or greater degree with other animal life forms, start with the simple reflexes. For a very readable purview of this topic is "Mama's Last Hug" by F. De Waal (2019).

The root of the word that gives the name to Part III is 'value'. In proposing the ordinal and cardinal order within related values, and for the assessments concerning unrelated or even contradictory values, one needs the mental functions that allows comparison of actualities and possibilities. Furthermore, the concept of value ties together all of the topics of philosophy that do not clearly belong to ontology or epistemology. But these two areas are necessary bases for the philosophic ideas that deal with the topics of this part of the book. The discussions of evaluations, have been developed to be consistent with the reality-based information and the knowledge derived therefrom.

What then distinguishes conscious evaluative mental activity from the rest of the mind is discussed in the Parts I & II? The short answer is that it is, by far, the most complex activity of the human brain-mind. Discussing it involves concepts that are either not covered, or are only of tangential interest to the questions concerning ontology and epistemology. This section introduces the commonalities and differences. How the various combinations of these particular emergent brain-mind abilities differ is the subject of the following chapters.

First, we will examine why only Model forming Living Things (MLTs) can have purpose, unlike the other two types of interactive things, the Reactive (RLTs) and the non-living (NLTs), both of which can only have propensities. To have purpose, one must be both alive and be able to conceive of a future. That is one of the characteristics of an MLT. (See § 10.9).

The underlying advantage, of the evolutionary development of greater evaluative processes, is the added value they provide in guiding the activities that prolong a life throughout the reproductive cycle. Such improved evaluation is especially useful in developing the best actions in a group setting, the natural setting for the genus Homo. Moreover, in humans it also allows, beyond sheer survival, the development of wisdom. The application of wisdom (how to best interact with the world) also involves the use of the will as discussed later.

9.13 The Ever-Changing Individual Brain-Mind

The science of embryology studies the transformation of a single cell, the zygote, created from two incomplete but complementary cells, the ova and sperm, into a complex organism. In the embryological development of our species, we can still see transient traces of our pre-mammalian ancestors, such as the gill-slits of fishes. The human brain also evolved by additions to and altering of the neurological systems that came before. The earlier systems are not totally gone, but rather were changed and/or overlaid by new systems. At the same time these earlier forms were integrated with the later systems. The earlier systems continue to organize and coordinate our physiological functions, reflexes and instincts. It is the newly expanded prefrontal cortical brain (and to some extent the lateral cerebellar structures) that provide the new cognitive abilities of the brain-mind that separates us most clearly from our closest living nonhuman relatives.

Philosophy, if it is to deal with the totality of the human experience, will need to search for wisdom in the context of our understanding of the evolution of human mental development. It also needs to incorporate the discovered provenance of these changes into our current concepts. To do this one will also need to use information gleaned from paleontology, anthropology, history, as well as infant and child psychology, as well as other pertinent areas of study. Infant and child development is especially accessible and instructive because most of the cognitive emergent properties of the human intellect emerge only with the postnatal growth and development of the human brain-mind in each generation. Therefore, the complete set of concepts concerning the higher evaluative functions can only fully apply to typical adult human beings. Furthermore, these developments in children and young adults can be studied in the here and now. These studies show that these advanced abilities are the result of the interplay of the many diverse forces that continue to mold the brain-mind throughout a lifetime. The brain-mind is not static. A book such as this can only point to some of the highlights of information that have come to the fore in all these areas, information that is being modified weekly – if not daily. This outline of the quantitatively enormous and complex emergent functions is meant to nudge mainstream philosophical thinking into using broader sources of information. This means going beyond the knowledge that could be ascertained from studying only (or even mainly) the output of historical philosophic thought.

These searches for understanding of the evaluative activities have often been limited to introspection and the experiential (not including experimental) observation of human activity. Until very recently, such limited information has been the basis of all the ideas and systems of philosophy discussed since the early Greeks and their contemporaries on other continents. What is needed is the evolution of philosophical concepts that will encompass new knowledge and understanding of the human condition. These studies are now underway, often by thinkers that started their work in other academic areas. Studying the history of philosophy alone is useful, but can only serve as a backdrop and base to that effort.

9.14 A Way Forward

A careful use of metaphors, by comparing the evolution of the physical aspects of life with the evolution of functional aspects of the brain-mind, can be useful. The study of metabolism deals with biochemistry and physical chemistry, physiology adds the physics of mechanics, energy transformations and feedback systems. The emergent properties of brain-mind are based on the levels of nervous system organization: the structures and relationships of the complex networks of neuronal tissues as they change over a lifetime. (See National Geographic, February 2014, and "Behave, 2018"). For a simple outline of the anatomy see the Appendix. The understanding of these newer properties must, however, still conform to the insights gleaned from the study of the lower levels of organization, even though the emergent properties of a brain-mind cannot be explained by the properties of less complex systems and functions alone.

A whole new set of concepts and terms needs to be continually developed to discuss these new properties. As Deleuze et al (1994) state "... philosophy is the art of forming, inventing, and fabricating concepts." Also, by mutual agreement, the import of specific classic words can and must be changed. For example, the word 'atom' persists, and is used in physics and chemistry, although the details from the initial concept of Democritus have changed. As new information became available the definition of this word/concept, as used in the basic sciences, had to change. The concepts concerning brain-mind functions have, for the most part, been historically developed in the fields of psychology, sociology and theology as well as philosophy. Before delving into such complex areas as ethics, it behooves one to formulate "corrected" vocabulary-concept pairs concerning these emergent brain-mind activities and products. This necessitates the adjustment of basic definitions concerning the characteristics involved. At a minimum one must determine what the necessary and sufficient characteristics of ethics (and the other terms) may be, and only then go on to differentiate subtypes as more characteristics are added or subtracted. Initially this amounts to creating a working definition for each of these concepts. At the very least such concepts must be consistent with the models available to us from careful introspection, close observation, and valid experiment. I will offer a start for aesthetics (Chapter XII) – arguably the 'metaphysical' concept area that has the least components in the realm of complex evaluations. The approach recommended above is used to provide an organizational basis, so that in the discussion of the components, we not only keep the properties of the necessary subcomponents in mind, but also how each of these contribute to, and underlie, the next levels of complexity.

Chapter X
The Concept of Value

10.1 The Cognitive Basis of Value

The perception of values can be shown to be based on the needs and purposes of MLTs. The perceived needs are based on real, objective needs of which there are several levels. These needs are shared by many RLTs, but only MLTs assign cognitively derived value to them. They are:

1* The needs of life functions, the classical biologic needs.

2* The needs that can satisfy the emotional/instinctual drives of the individual organism.

3* The needs that support the success of the group (community) in a social species.

4* The needs required for a stable environment that can support the first three needs. This is the need to preserve and replenish, if possible, all natural resources for future generations. (Schumacher, 1989).

Value is based on a relationship between Need and its Satisfier(s). The actual and perceived magnitude of the (V) of satisfier (S) is derived from the urgency of need (N) and inversely related to availability of S, the specific condition or material that can satisfy the specified need. This can be written in a simple relationship:

$$V \sim \frac{N}{S}$$

This relationship is the basis of what is called extrinsic value. An intrinsic value would only apply if the satisfier is the same as the source of a need. In other words, the value of the self – for the self.

The urgency of the need depends on how close the actual situation is to the limits within which the organism, its mental stability, its group and its environment can function and maintain itself. The availability of resolutions may sometimes be met by a single item, such as water for dehydration. But most often it is a set of such items or conditions. There are two types of sets. The first is when individual items in the set can resolve the need, but have different abilities in this regard. The second is more complex, and is composed of items that affect each other. An example would be the blood sugar levels of an organism. The availability of glucose and insulin need to be balanced against each other to arrive at sugar levels that are in the life supporting range. This is in fact supported by three systems. The neurologic by signaling hunger or satiety, and the automatic metabolic and endocrine systems.

The human brain-mind is capable of deriving relative values by processing the signals coming from the internal sensors of hunger, thirst and pain among the many needs of biologic importance. Our NS also automatically manages the needs for short term social and environmental stability. However, the long-term needs in all these cases requires an analysis of the possible models of the future. This can only be done consciously, although unconscious and preconscious processes can feed information to the conscious processes. As Table II, at the end of § 9.11 shows, this is a relatively slow, difficult and fraught process due to the many variables involved.

Consider this thought process and recall that we can give the value of symbols cardinal numerical form that approximate the sum of quantum effects. But, without some standard measurement available, we are more likely to use relative ordinal numbers that express subjective evaluation.

Once we have given relative numerical values to the symbols, we calculate the following relationships. Let us say the need is a constant of 10, but the availability of its resolution is 100, 50, 10 or 1. The values (of the things or actions that satisfy the need) would then be 0.1, 0.5, 1, and 10. If the value is much greater than 1, competition for the resource will be predominant. If the value is much less than 1, then a laissez-faire approach of acquisition will rarely harm anyone. When the value is near 1, cooperation may become a significant factor for either decreasing the need or increasing the availability of its resolution. Complications will arise because the actual needs and resources change over time, place, and circumstance. Because these three responses are possible, ethics may be called upon in an attempt to balance them against some created norm. That is the logical conclusion in facing social life's exigencies.

The exact value for such needs as the amount of water, food, shelter is at best an approximation. Also, there is actually a range of availability to which our metabolism and bodily functions can adjust. Therefore, the perceived range of the need(s) and the availability of the various real resolution possibilities are usually judged on an ordinal scale for any particular circumstance. "I am thirstier than I am hungry".

However, the instinctually and emotionally based values and preferences do not always conform to the logic of such direct physical relationships. Mental models created by confabulation or fantasy add to the large range commonly seen in these I-E values. When no conscious reference is made to models that are based on confirmable reality, then these emotional values are perforce prejudgments. Giving ordinal values to competing personal emotions is therefore more complex than measuring the physiologic ones. The psycho-social needs are both hard wired in neurodevelopment as well as modified by subconscious adaption to social pressures and examples. This makes the subsequent conscious, prefrontal cortical assessment of both needs and possible resolutions much more difficult and at times impossible. Hence the questions, "Why did I/he/she do that? What need drove them and why was that particular action used as a possible resolution?" Only a study of the complexities of the NS will give us an inkling. There is some hope for more accurate answers now that we can "watch" the brain in action with fMRIs (See glossary) and other real time studies of the brain-mind in action. But they will always be only more accurate, never absolutely definitive.

The instincts and emotional reactions are based, by the forces that affect evolution, on patterns of behavior that supplied the needs of our forebears in the temporally distant and often very different physical environment. Hence, the current value of an instinct is related to how closely a current environment mirrors the past. A person in a non-industrial, rural environment is more likely to benefit from instinctual values than a person confined to a modern crowded city. The instincts may even be counterproductive and antisocial in the latter case.

The conflict between the rational and instinctive value systems is one that the field of philosophy deals with better than a purely scientific one. Wise decisions weigh both confirmable information and the limits imposed by strong instinctive and emotional imperatives. Wisdom firstly acknowledges that consciously made decisions can rarely be described in absolute terms, because information available to cognition is usually incomplete. Then the seeming clarity of emotional and instinctual responses must be questioned, because these urges may not fit the current scenario of actual needs and resources.

Because of the great advance in understanding provided by modern science, especially the neurosciences, including psychology and sociology, such information should be incorporated into current models of value. Part IV contains the chapters that provide attempts at such model building.

10.2 Preferences – the I-E Values

We all have preferences on which we base many choices. Preferences drive the choices which are the least based on cognitive evaluation. We speak of tastes, likes, dislikes, distrusts and fears in this regard. All these are shared by many animals. These are values that are most often formed by un- and pre-consciously formed evaluations. It is in the field of Aesthetics that we find the best discussions of these kinds of evaluation. These values are heavily influenced by the instincts and emotions, but experience and associations can play a large part, including beyond early childhood.

At their best, the models used in discussing morals, ethics, and justice should have, and usually do have, a large input from the cognitive system. However, preferences arising from C-L thought are commonly different than those arising solely from I-E systems. The former has the potential of being rationally and logically discussed in the light of shared or conflicting mental models and defined facts. A logical discussion of pure emotional preferences is limited to the examination of the workings of the non-conscious systems in which they arise and to determine the consequences thereof. Beyond that one can only say, "I like\hate that" without logically defensible material justification. C-L cognition, as used above, is not the same as in Kant's "Critique of Pure Reason "(1781) (as per Wikipedia entry of same name (2019), "I do not mean by this a critique of books and systems, but of the faculty of reason in general, in respect of all knowledge after which it may strive independently of all experience." Independently of all would mean no

The Concept of Value 165

basis in sensory experience — an impossibility for an adult human brain-mind. Even a newborn has had some sensory experience during the prenatal period. The emotions and instincts in a newborn infant, are not thought to be consciously evaluated. This is supported by the later slow development of reason. The I-E system are their only sources of action. But, within a few months and years, sensory experience will inform the creation of cognitive models that have also been influenced by the inborn intra-nervous system responses to that same experience. These are the abstractive and then the subsequent imaginative processes discussed in the first parts of this book. Neither process alone can serve the full needs of a typical human adult (or even a young child) when determining the best models possible for promoting the thriving of that individual in its environment.

10.3 Value of Life: The Process Versus a Specific Instantiation

The value of life as a process is very high, if not paramount, once the basic P-laws are accepted as a given. This process is the source of the most potential and variety at the spacial level in which living cells occur. Without life there would be no philosophy, ethics, right or wrong – nor beauty, love or mercy – and their opposites. It would be a relatively simple, and sterile, universe.

The value of an individual living being is variable. It may range from the perceived value of near zero (for a single bacterial cell) to uncalculatable or near infinite (by definition) for a human being. The unimportance of a single bacterium is based on the ground that it is but one example of a myriad of its genetically identical clone-mates. But what if it was the last of its kind?

The genetic and experiential uniqueness of that group of clones (from an original zygote) that make up each human being, compounded by the unknowable potential meaning of each, leads to the signification of a high but unmeasurable value. However, in sum total, the ascendancy of man since the first agricultural revolution (~10,000 BCE) has led to a decrease in the overall diversity of life on earth. This loss of balance, due to human overpopulation, is also a major underlying factor to the current climatic changes. The energy consumption of one human, based mainly on fossil fuels, multiplied by the current 7.5 billion (and increasing) individuals has led to the current situation.

How is an ethical system to be devised and evaluated with these parameters in mind? How to weigh the potentialities of an individual against the planet wide harm of this anthropogenic era?

Deontological, that is rules based, ethics have no problem with this seeming conundrum. "The rules are the rules" and the consequences are of no consequence in these ethical systems. But this ignores the underlying, and unacknowledged, assumptions of a transcendental utilitarianism. Religions only claim that the purpose of the rules is that of the gods. It is also a stance that disallows any reality-based evaluations that preclude the inclusion of fully imaginary axioms. Can a realism based, partially measurable – and therefore hypognostic – ethics be any more convincing? A realism that acknowledges the importance of not only material consequences, but also the psychologic consequences to organism with complex brain-minds. The realist stance includes:

1* Our models of the effects of the axioms of physics (P-laws) are correct at the level of life.

2* The maintenance of an environment leading to a long term maximalization of the variety offered by the process of life is most important. This is the goal.

3* The human species is, by necessity, communal.

    3a* To achieve #2, cooperation, at the communal level, is on average more successful than selfishness to avoid the "Tragedy of the commons".

4* The actions of individual human beings are driven by two main brain-mind processes: Instinctual-Emotive (I-E) and Conscious-Logical (C-L).

4a* Before action, C-L oversight of I-E impulses leads to a higher achievability of #2.

5* Ethics, the rule of thumb (initial) rules of action, are developed as a guidance toward the means leading to #2*.

A good place to start are the current rules-of-thumb, and then place them under the microscope of these five assumptions and their correlates. A further effort should be made to order the rules according to the consilience with agreed upon needs. This will be a long and arduous task, not to mention a contentious process. But if it is not undertaken the future looks bleak.

10.4 Identity

Identity has been a subject of philosophy, as in the tale of the two ships of Theseus, discussed since at least 400-500 BCE. In short, the story is that of a boat that needed refurbishment. As the pieces were removed and laid aside, new replacements were added. Over time, those laid-aside pieces were rejoined, thereby reconstructing the original boat. Then the question is posed – is either boat identical to the original – and if so which one.

The answer entails further discussion of what the concept of identity entails. The narrowest meaning would be the description of an object or action at one instance of time. This is because identity would also entail all the relationships the object has with, or the action it imparts on other matter/energy in space, and over subsequent time. The most important interactions include those with the various energies such as gravity, electromagnetic and so forth.

This is too restrictive for general use, but is included to underline the relativity of identity. For most purposes three characteristics need to be compared: the material, functional and relational identities, alone or in combination. If we wish to meet all three criteria, then the narrow definition would apply. Under which circumstance are the individual, or combined, characteristics most applicable? That is why is the question of identity even posed.

The reassembled ship is, materially (almost) identical to the original. Almost, because there is bound to be a splinter or two missing, a nail or two brand-new. Both ships are about equally functionally identical in terms of carrying capacity, draft and other similar properties. The least comparable will be the relationships to many outside objects, and factors – such as the carpenters and shipwrights, length of foreseen service and perhaps the name on the prow. There are other relationships, such as emotional affinity of the crews (we need two sets now) and ownership which also affect which one may be seen as the original vessel of Theseus. Much of Aristotle's discussion centered on the causes, final etc. (§ 4.7). Causes fit into the category of relationships, which makes questions of relational identity or sameness more difficult to evaluate.

In a two-dimensional analysis of materiality and functionality we have 4 possibilities. The combination (Combo-1) of new function and new material is the least identical. Old (original) material and old function is the most identical (Combo-2). The other two combinations lead to limited material or functional identity (Combo-3 and 4 respectively). Trying to add the third factor, relationships, is a fraught exercise. The possible combinations are myriad as are the levels of identity. Usually, the best one can state are ordinal levels of identities. An exception is when only the relationship is used as a factor, such as siblingship. So, what is meant be a phrase such as "all men are inherently equal, that is identical?" (See next section).

What of personal identity? The first thing is to consider the passage of time. The more time between two related entities the further we move from Combo-2 (old-old) and move toward Combo-1 (new-new). Furthermore, there are differences between material identity and psychological identity.

In terms of the atoms a person is composed of, with time they are almost all new. But structurally, that is the overall anatomy at both the macroscopic and microscopic level there are lesser changes. Aging, accidents and disease are the major contributors to the level of change. However, our bodily anatomy is usually of less importance in our perception of identity than psychologic changes due to brain-mind changes. All the same factors pertain, but are magnified because the perception itself is a function of the brain-mind. The largest changes that we notice are in memory, motor abilities and social interrelationships. The latter are the most changeable because they involve the change over time in more than one person. The conclusion is that no one is the same, but only similar in structure, function and relationships over time. It is the amount of difference that is important. The greatest change is of course the death of an individual. (See § 15.6 Life and Death). But before that the changes that affect the brain-mind are the most troubling – especially in the areas of self-care and social interrelationships.

In summary, we all change over time. What we call self-identity, seen either from one's own perspective or that of others, is the usually gradual, imperceptible changes – the similarities in all major functions. When the changes become noticeable, we are indeed a different self, albeit related to all the old selves by the processes of cause and effect. Short of death, the previously explained thought experiment of the vat full of cells(§ 2.1) is a good place to continue the discussion.

10.5 The "Equal" Value of a Living Thing
"All men are created equal" is an enlightenment ideal enshrined in the Declaration of Independence. The whole quote is the ethical foundation of the legal structure of the United States of America. To wit:

"We hold these truths to be self-evident, that all men are created equal, that they are endowed by their Creator with certain unalienable Rights, that among these are Life, Liberty and the Pursuit of Happiness.—That to secure

these rights, Governments are instituted among Men, deriving their just powers from the consent of the governed."

This 18th century document is part of the mutually supportive structure of much of western law and common ethics. But it also has been a source of disagreement in the discussion of ethics for several reasons, briefly outlined here.

Firstly, are the propositions truly self-evident? I think not. From a realist viewpoint, self-evidence only indicates mutual agreement, either logical, emotional or both. The most obvious part today is the proposed self-evidence of a Creator at that time. Taken as an axiom, it is a part of transcendent theology and philosophy. The reality of such a transcendent being is highly controversial, entailing competitive and in many ways internally incompatible world views. It should be no surprise that ethical systems based on incompatible basic axioms concerning the underlying realities of our universe are at best only partially overlapping. This is especially true when considering the ex-nihilo creation of the world by a god compared to the hypognostic statement of the (currently) unknowability of the state of the universe before the Big Bang and the subsequent, partially understood, cause-effects as determined by the P-laws.

The same can be said when it comes to the source of "rights". In the first case, they are sourced in the transcendent, in the second they are the consequence of temporary agreement as codified in written laws or transmitted by the various subcultures that comprise human groups. When gods are involved, the primary disagreements will be which gods underwrite which rules. This is just as relative as the many ethical and legal systems created by humans over the ages. This should not be surprising, because the concept, god, is uniquely understood by each brain-mind. All philosophical discussion must therefore first decide whether it is theological at its base or founded in a hypognostic realism.

If rights are not given by a god or a community, what are they? I believe the axiomatic automaticity of rights is a great mistake. If analyzed closely, we can see that they are the acknowledgement, firstly, of the physical and psychologic needs of humans and other beings. They are an expression of justice in that they attempt to codify certain needs as absolute. Rights can be taken away – that is they can be ignored if the subject of the rights is found to be of low value to the community or an individual who previously agreed to conform their actions to uphold the specific right. The needs remain. The interrelationship of needs, value and availability of its satisfiers (resources) was dealt with in § 10.1 above.

Happiness as a goal will be discussed in §§ 11.6-7, and found to be a Promethean quest. Liberty or freedom is closely tied to relationships between the self (or a defined group) and some defined environmental conditions. The concepts posit total self-determination, and non-interaction. Since the interaction of all matter/energy is only a question of degree, not an absolute condition, the level of interaction must be understood and the significance evaluated. Agreement can be found, but it will never be absolute at the margins, especially in unusual, rare or unique cases.

The right to life is, from a naturalistic perspective, an oxymoron. Nature only provides the possibility of the life process, not its guarantee. This is as true for each individual cell and clonal group of cells (that is the individual(s) that share the same DNA), as it is for the process. Agreement for the human-given rights, can be found in the latter case only by assessing the relative value, of a specific cell or multicellular clonal aggregate (such as a plant or animal) against the value of an ultimate goal, such as the proposed fostering of the general process of life and all that entails. Other ultimate goals can be proposed and defended, such as the Agathonism of Bunge (2003). But ultimately one goal must take precedence for a universal ethic to evolve. Only by finding agreement in this conceptual arena can the dependent values be examined further. Hence the never-ending Promethean philosophical task of increasing understanding and wisdom. This effort is necessary, to develop workable ethical systems, considering the limiting hypognostic state of brain-minds. Promethean, but hopeful of progress. Let us continue this work.

10.6 The Trolley Problem – a Popular Thought Experiment

This is a thought experiment meant to compare utilitarian vs deontological (normative) Kantian ethics. Its best-known version is that put forth by Phillippa Foot (Wikipedia). It goes as follows: You are standing near a trolley track watching an out-of-control trolley is speeding toward a group of five persons who cannot (for unstated reasons) get out of the way. You have the ability to divert the vehicle, by throwing a switch, to a side track on which only one person is in the path. The question is – Do you throw the switch to change the direction of the trolley? Many subsequent versions of this dilemma were published, including pushing a large person into the path of the trolley to stop (derail?) it. All versions have been more useful in studying moral psychology rather than the clarifying the "best" goal for an ethical system. It does not offer a path to an answer as to how to value different outcomes. For a readable overview see Cathcart (2013).

It does however point out the unresolved problem of the value of personhood, which centers on how to define a person. This is a question that bedevils the fraught subjects from abortion and euthanasia to capital punishment and war. Each of those topics warrant its own book, but here I offer some thoughts that might help the discussions along. These address what defines a person, what is their value, and finally the ethics of each (see Chapter XV).

Firstly, any person is a structured system of individual cells that are clones of the original zygote produced by the combination of a sperm and an egg. Usually there is only one clonal group of each kind, although identical twins are an important genetic exception.

The twins are normally seen as two distinct persons. But there is another, much rarer problem of identifying personhood, that of conjoined twins. I have personally (as a physician) treated such a pair. Their bodies were fused at the level of mid-spine and as a consequence had two heads and four arms but only two kidneys, one bladder and two legs. How many persons is this unfortunate, inseparable pair? If there are two, which one "owns" the kidneys, the legs? Which brain-mind gets to choose where the two, independently controlled, legs go? In practice, the two brain-minds had to verbally discuss where they would go next – and then coordinate their actions. This seems to tell us that it is the brain-mind unit that is most central to personhood. Just as in the typical identical twins. None of the other organs meet that criterion. (This is also confirmed by how we define an organ that has been transplanted – it now "belongs" to a different person).

I will reiterate the thought experiment of §2.1, which I believe is novel. Let us believe that there is a method where a typical person is placed in a large vat containing a solution which will disassociate all the living clonal cells, without killing them, and then supports them so they continue to live – as individual cells. Although related, this should not be confused with the "Brain in a vat", an updated version of René Descartes's evil demon in a thought experiment originated by Gilbert Harman (See Wikipedia article for details.)

Is this group of cells still a person? Is the person dead? By losing the structure, it only has the tropisms of unicellular life left. The other biophysical properties that emerged from the structured growth of the zygote, by cell division and differentiation, are gone. All the emergent properties of a brain-mind such as thought, memory and the psychologic, social properties that the person had are also gone. Is the person dead? The clonal cells are all there and alive. But all the properties that distinguish a person from a group of amoebae are gone.

This points out that death, most clearly, is the disruption of the properties of a cell. All other changes to the structures and functions of organs and organisms, made up of the clonal cells that form them, deserve a different concept, and name to address the thought experiment of the vat of clonal cells. Another option is to speak of the extinction of a clonal group, when all its cells have died. This is what happens when the physiology of a person disfunctions to the point of total clonal extinction. Then the person is also extinct – or in common parlance – dead.

But that leaves the question- which of the emergent properties that separate a person from an amoeba are necessary and sufficient to make the distinction? Firstly, the basic biochemistry (and DNA) of all mammals is ~ 90% the same. A start can be made by differentiating the persons components into the physiological necessary supporting organs, surface characteristics, and the properties of the human brain-mind. This is how we, in practice, differentiate a person from any other living species, or as belonging to a different ethnic group. Some organs can be transplanted (interchanged) from a non-human animal. Demoting persons of a different ethnicity to a non-person status is fortunately a thing of the past for any typical educated person. The functioning brain-mind is therefore the best candidate of a necessary and sufficient difference that defines an organism as a person. It does not however, answer the question of how complete the properties of that nervous system activity must be. What then of the value of a person?

For a broader discussion of value, you may preview §§ 13.2, 13.3, but in the context of ethics Kantian and Utilitarian camps predominate. Kant indicates a person should never be used as means but only seen as an end. It is a "the means justify the end" approach rather than the extreme Machiavellian "the end justifies the means" position. However, the Kantian view, invites paralysis of action (the means) when persons compete for a very limited vital life resource. How does one choose who benefits and who suffers or becomes extinct? Is laissez-faire, by not choosing means, the only option? How does one choose between two ends (persons)? Furthermore, if persons are, in essence, of infinite value a paradox arises. Infinity (∞) is a fraught concept. We know that (∞) plus (∞) = (∞), as does (∞) times (∞). Not more nor less. In that case the value of one person equals that of five — or a whole human population of the world! So, the death or losses of one equals the death or losses of however many. In that case no resolvable ethical questions of harm or helpfulness pertain.

This is where utility creeps in. If a person has a non-infinite value, it is a reflection of how that person meets some need. The need can be that of an individual, a group, or the living environment as a whole. But the need itself supposes a choice between outcomes – the ends. Here there is no paradox of "the infinite value of competing, undistinguishable ends justify some ideal means", because there is no meaningful; difference between the ends. Instead, we have to find a way to assign ordinal value to the ends. But this can only be done if there is only one other ultimate end. Without this no clear ethical guide can be developed.

But there is a further grey zone. A living clonal human being, that has none of the characteristics that separate personhood from animalhood – is incomplete as a person – and needs a system of ethics that acknowledges that difference. Without that one cannot proceed on a C-L basis, and is left with only I-E evaluations. This is enough in the ordinary cases – but not in the exceptions that beg for clearer guidance to the question – what should one do? Only an ethic that includes both brain-mind evaluations will satisfy the inquiring mind. And it will only satisfy a particular situation, because there can be only incomplete comparison of value of proposed means.

10.7 "Pure" I-E goals?

There is however another complicating bidirectional interaction. The desired outcomes, the ultimate consciously chosen valued goals, are also input for, and not only the result of, recursive preconscious evaluations. The neurologically derived preconscious goals can have a similar mental function as axioms, in that they can act as givens in our calculations. Although they are only marginally based on rational evaluations, and mostly or completely on the reflexive responses, they are often just as, or more, influential. These unconscious evaluations drive actions as surely as purely rational conscious evaluations such as mathematics and logic. This may frustrate the application of the latter in handling difficult problems.

The various pre- and un-conscious subsystems can also be in conflict. An example would be saving oneself versus one's infant from an immediate, non-human threat to both. The instinct of a prairie hen to act injured and lure away a predator from the nest shows that "deception" does not have to be conscious or aforethought. As such these "decisions" lie outside of ethics. A purely rational, conscious decision in such a case is rarely possible for us non-Vulcans, of Star Trek fame, especially when little time passes between the stimulus and the action.

Further common examples are the instinctual responses which are already inherent in infant animals. That animals have such instincts is now accepted, even though the importance of I-E responses has been often minimized for human behavior. Daniel Kahneman (2011) divides brain functions into a precognitive System 1, versus the cognitive System 2. I prefer to use the term "I-E brain-mind" for the former, as discussed in § 4.8. The emotional system is but the most recent part of this non-conscious brain-mind complex. These innate evaluations are predetermined models of reaction based on genetically based structures and functions, created by the feedback loops of evolution. As discussed in § 4.9, instinct is the mechanism driving the complex reactions in all animals that have more than the simple motor reflex capabilities. However, the difference between motor reflex and instinct is one of complexity, not kind. As a human infant develops, small, individual-specific inputs from its environmental history begin to change the reactions of its purely reflexive brain-mind. Eventually, instinct can be overridden by habits (Skinnerian conditioning can be interpreted as such) as well as by strong input from the cognitive system, as also discussed by Boyd (2016). Alone, reflexes and instincts cannot give rise to purposes (chosen goals) or morals and ethics. What is needed for the emergence of these projective evaluations is a nervous system that can imagine a future.

10.8 Emergence of the Concept "Future"

So far, this book has mainly given examples of emergence from the sub-atomic to the instinctive brain-mind. This overview stressed the centrality and necessary possibilities inherent in physical emergence. This served to underpin the idea that atoms and brain-minds are connected through many levels of increasing complexity. Each level added new possibilities of interaction. The evolving increase in the complexity of primate brain-minds eventually led to the mental ability in early hominids to imagine an extended future. It allowed a release from the constraints of automatism in the form of motor reflexes through instincts and emotions. It is of interest that the great apes can show, by their actions, (De Waal 2019), that they are able to plan for some near future (±24 hours) goal. The H. sapiens line of descent evolved a further expansion of this trait than their ape cousins did.

The result was the ability to plan actions and at the same time to imagine the outcome of those actions on the self and the greater environment. This evolutionary outcome was necessary for the advent of the complex mental evaluations needed to create models of values, including the values of social and environmental ethics. This ability coevolved with the concepts of right and wrong, helpful and evil intent, justice and revenge. (See § 13.0) All these concepts tie the memory of the past and the impressions of the present to the yet-to-be imagined future. Without the benefit of very recent, science-based understanding of the capabilities and limitations of the modern human brain-mind, these concepts were commonly thought to emanate from transcendent sources and to exist independently of human conceptual development. Although our inherent hypognostic state allows for the possibility of other, wiser, more powerful beings (beings up to and including the concepts inherent in imagining gods), their proposed intervention can now be seen as unnecessary for ethical concepts to exist. And, because these concepts are the creation of human brain-minds, they are necessarily open to incompleteness, contradiction and misunderstanding. It is one of the tasks of philosophy to rationally improve these concepts so that they are more complete, less contradictory and lend themselves to better mutual understanding.

This process will involve the creation of models whose formulation depends on an understanding of the evolved changes from simpler nervous systems. These changes allowed the laying down of, and the retrieval of, personal memory among other abilities. The creation of new reality-based models is (will be) based on a variable human ability. Such philosophically useful constructs are the formulations of typical 25-70+ year old human adults, as discussed at the end of § 4.3.

10.9 Beyond the Preconscious Adaptive Mechanisms (PAMs)

The PAMs have previously been defined as all the preconscious activities of a brain-mind other than the emotions. It overlaps with the I-E system which does not include motor reflexes and habits but includes the instincts.

The ability to consciously compare and evaluate different possible actions for the purpose of meeting needs is the most recent emergent capability of the evolved human brain-mind. Thereby one can derive the ordinal values of things, states and processes. This ability is of course most notable in a typical adult H. sapiens. These conscious comparisons and evaluations lead to plans of action. These may then be evaluated further by imagining the possible results on meeting the needs of other beings and one's own future self. This can then give rise to ethical guidelines and wisdom. It is not however a guaranteed outcome. The focus of the mind must go beyond egotopian planning if these social and environmental consequences are to be considered. This is often neglected because the conscious mind does not process more than one problem at a time, although rapid switching between subject matters is possible. However, adding more layers of future outcomes with multiple potential effects, takes time, effort and some empathy.

10.10 Conscious Evaluations Are Serial

To make comparisons for planning, one must have, at the very least, short-term memory and consciousness. Such mental juxtapositions are serial, recursive, internal dialectics between the related mental models that are created by the imaginative processes. The speed with which we think makes the process of creating and retrieving these models seem simultaneous in time. However, reflection on one's own thought processes will show that although many other processes are running in parallel (for example seeing and hearing at the same time), the specific process of comparison is serial in nature. This limitation gives rise to the concept of focus. If one's conscious thinking is not focused, one cannot create distinct models, but only vague impressions. Most everyday thoughts deal with more or less vague impressions that only incorporate the common and typical attributes of any one concept. In dealing with the here and now of the environment, that is all that is possible within the physical constraints inherent in the abilities of the brain-mind. Building more accurate models takes more time, more information and recursive comparisons of the various models under consideration. But these more accurate models are exactly what one is searching for in philosophy and in the sciences. Once these models have been created, they may replace the vaguer ones in memory by creating newer memories and thereby provide a newer (hopefully improved) and expanded mental database. This leads to an increased probability of developing an ever more accurate worldview — one of the aims of wisdom. These time-consuming imaginative processes can also take place subconsciously during wakefulness or sleep and are subsumed under the term's insight and premonition. It is the process that led to the "Eureka" of Archimedes, known colloquially as the "aha moment".

10.11 The Parallel Evaluations

Truly parallel comparisons would only be possible if there were separate circuits that can create different models based on the same information during the same timeframe. The closest examples we have to this, is the parallel workings of the PAMs and the higher conscious circuits, all reacting to inputs from the abstractive systems. Complex motor reflexes may have parallel components, especially on the output side. Fully parallel systems within the emotional or cognitive system do not seem to exist. Serial, recursive, and multi-branched processing is the norm. In any case, the systems of the brain-mind operate at the level of milliseconds (one thousandth second) and the results are only available to us for active logical comparison as they reach consciousness.

In summary, consciousness, emotion and instincts are part of the outputs of the human brain-mind. Their emergence was made possible by the ever-increasing complexity of nervous systems' architecture and chemistry. In comparison, the earliest forms of NSs controlled something as 'simple' as the reactive reflexes, such as the coordination of jellyfish motility, the writhing of an earthworm uncovered from under a pile of leaves and the flight pathways butterflies take as they flitter from flower to flower in a gentle breeze.

A more thorough discussion of the evolution of nervous systems through fish, amphibians, reptiles, early mammals, and later primates and finally early humanity are left perforce to other books and authors. We will continue our discussion with the inferred, but necessary, properties of our remote, but human, ancestors.

Chapter XI
The Function of Purpose

11.1 The Expansion of Imagination and Memory

The early hominid species shared, as do we, the basic abstractive system, reflexes, instincts and basic emotions of their primate ancestors. The systems were not so much replaced as added to by the increased functions of the prefrontal cortex (PFC) – the expanded brain structure of the hominids. Imagination, the process that creates new concept, is dependent on the emergent properties of these areas as well as the increases in memory.

This developing structure, which physically sits above and in front of the others, is highly connected to and functionally interactive with the earlier systems. This expansion of cognitive "thinking" could and did supersede the reactive outputs of the lower systems in many situations. The documented growth of the cortical structures over succeeding ages is the physical sign of this development as described by the mutually supporting data developed by research in paleontology, archeology, embryology, child development and other areas of research.

11.2 Universal and Personal Goals

A search for universally held goals and therefore ethics, is often muddied by a desire for the primacy of one's own vision of perfection. This process is easily dominated by one's own egotopian visions, over all the hopes, dreams, goals, aspirations and egotopias of others. The desire for that preeminence, to be the "alpha" of a group, whether physically and/or cognitively, is one basis of all social conflict. Only when a vision is shared by a small, utopian group can conflict be avoided for a time. The group must be quite small to forestall variations in detail because the details must be understood and agreed to by all involved. The prototype for such detail would be the rules of any complex game. Nevertheless, this kind of agreement on values cannot hold for long, even in a small group, for one basic reason – the impermanency of a human idea. Games too see changes in rules over time, often giving rise to competing variants. This changeability of desirable goals is unavoidable because goals are the product of a brain-mind that continually has new inputs. A brain-mind state cannot be a static condition, because the chemical and physical states of a brain are constantly changing. To be sure the changes are usually incremental and relatively minor. Usually, there is some coherence with the previous goals because newer ideas are also influenced, perforce, by those held before. As a result, the chances of two minds changing, over time, in total concert are zero. The possibility of a complete set of universally held goals is therefore also zero.

11.3 A Short Discussion of the Limits of Near-Common Universal Purposes

What is, can or should be the purpose of one's life? This is closely related to what the meaning of one's life might be, as discussed below. These questions have been asked by the philosophically minded for millennia. As noted above, the aim of any project is to move toward a goal by a set of means. Aims include all the foreseen steps along the way to the conceived endpoint. To be the best means they must be consilient with each other and reality. This excludes ideals, reserving this term for imagined goals which are unsupported, unsupportable or even impossible when tested against our current understanding of reality. The goal driven paths taken in the process of living, that is the purposeful actions which are consciously driven, are limited in their scope by the inherent physical possibilities, but are not dictated thereby in their details. These universal limitations are in addition to boundary setting aspects determined by the PAMs and their interactions. They are furthermore limited by our imaginative power. Finally, the maximum possible number of means greatly exceeds those that could be developed by any one brain-mind.

The details of the goal oriented human life are not dictated a priory (See the discussion of causation in the Ontology § 4.6), but vary according to context, ability and resources. "There are many ways to skin a cat" is a rather gruesome way to make that point. That is exactly the point of ethics. Once given a specific goal, there are many ways to reach it – but – the paths may not be equally desirable in an ethical sense. The means to an end, are also under the purview of ethics. In everyday living, most personal means are chosen as much by emotion as logic. The a priory values promoted by the emotions are outside the purview of a rational ethics. One can however assess the impact of the emotions on one's ethics, and how those emotionally colored values can be influenced and modified by the conscious systems. Kahneman (2011) offers many examples in this area.

The conscious, achievable purposes of life are therefore limited by the realities of our environment and the abilities of our brain mind. The meaning of life derives from, and is, all the consequences of one's actions and inactions. The understanding of that meaning is limited by our incomplete grasp of all the actual consequences of past and future actions. We can never have a complete and accurate picture. It should be humbling to realize that all energy transfer (actions), from the smallest quantum of energy exerted, will have meaning, however insignificant, or potentially great. An avalanche is but the sum of many tiny snowflakes. Who is to say which one was the most important?

11.4 What Are Achievable Personal Purposes?

What kind of outcome can be expected when developing plans for the future of the self, a family, a community and mankind? The answer seems to be those purposes, whether specific or broad in scope, will be somewhat nebulous in content. Limits of the possible are inherent as set by the P-laws of classic ontology as well as all the variations allowed by the emergence of new possibilities based on the complexities of living things, especially the MLTs.

The possibilities within these limits are seemingly infinite to the human imagination. An understanding of the real limits concerning the range of near-universal goals, and the infinite variations allowable within these limits is necessary. This will underlie the following discussion of such goals. Evaluations of the many variations can then be undertaken in perpetuity.

11.5 The Ethical Purpose of a Life

To answer such questions as "What is the purpose of life?" one needs to define purpose. What is meant by life is defined in § 3.3 and reiterated in the Glossary.

Before we define 'purpose' in this context, I wish to exclude those kinds of goals that are anethical. By anethical I mean a proposed goal and its means of achievement that have no value in an ethical sense. Such endeavors have, by definition, neither helpful nor harmful consequences relative to an ethical value system. For example, the use of 'purpose' in the statement "one purpose of a hammer is for hitting nails". It points to a materially instantiated tool whose primary meaning is in the consequences of its designed-for use. These leaves the ethical and unethical purposes. By the latter I mean a proposed goal or actual willed action whose effects are harmful, directly or indirectly, to living things. It speaks to the expected consequences – not the unforeseen. In the first case one can speak of mechanical purpose versus life-affecting purpose. With that background I propose this description Ethical Purpose:

An imagined future state which may or may not be consistent with one's personal morals. Each ethical purpose (goal) is often based on the perceived needs of life in general, as understood consciously and emotionally. Secondarily, such purposes, in a living context, are then open to ethical value judgements. A broad group of personal means defines an Egotopia. A Utopia consists of ideal states of being that incorporate the consilient purposes of a defined group. The value judgement may be positive (ethical) or negative (unethical) for each foreseen consequence. The ethics of the overall meaning (consequence) of the goal is then arrived at by deriving a weighted average of all the effects. (As discussed in § 9.9). This judgement must be understood to be possibly in error because of all the unknowns inherent in the complexity of interacting living beings.

A short foray into a thesaurus indicates that some of the non-tautological synonyms for the word, purpose, are: aim, aspiration, and target. Each of these terms needs to have a clear, if nuanced, definition to avoid nonsensical conclusions based on using them as tautologies in some instances and separate concepts in others. In this book I have used them in the following way.

By purpose, goal or end, I mean an imagined and desired final result. This finality is judged against the certain end of one's own life and all the foreseeable consequences thereof.

By aims I will mean that subset of imagined events that are derived by a clearly definable awareness of imagined intermediate results that are temporally and causally necessary between the present condition and a more distant (in time) end.

The final end and intermediate aims are easily confused. Our minds cannot process, or for that matter know (create accurate models of), all the necessary inputs needed to achieve any desired complex future reality. The state of the current reality is obviously possible (has a probability of 1.0) and has resulted from the interactions of all the basic matter, energy flows and emergent properties that were necessary for it to come into existence. One should recall that all emergent properties arise out of the organization of the first two components. In order to make mental models that are adequately accurate in their predictions (for the creation of realistic means) we must include the results of interactions of any one component of reality with all possible others. To do this we depend on the axiomatic cause-effect analysis of each of the proximate causes (a combination of the formal and efficient cause of Aristotle) and their possible effects. When we name these proximate causes, we do so with the intellectual understanding that we can support their meaning by the innumerable cause-effect chains back to the P-laws, which, along with matter/energy in all its forms and interactions, are the ultimate causes. Of the pre-Big-Bang causes we can have no rational models at the present because the emergent post-Big-Bang P-laws don't seem applicable. We can only base our best models on current, scientifically derived data and our own mental experience.

Mental experience includes all stimuli to the brain-mind as sourced externally and internally to itself. Experience feeds into the creation of a model using the results of personal external historical events as well as one's own imagination and confabulation, whether interpreted as fact, fantasy or revelation. Revelation can be defined as insight attributed to the intervention of transcendent, unknowable forces that thereby create novel unexpected concepts. Secular insight is attributed to the workings of the subconscious mind, denying such otherworldly input. Revelation, by definition, is not testable, predictable or foreseeable. Revelation is dependent on the history of the individual causes and effects of particular brain-mind. These causes and the subsequent conclusions cannot, at this time, be validated by scientific or any other confirmable means. Based on current advances, the causes of secular insight may be realistically discoverable with further advances in the neurosciences. However, the complexity of the brain mind makes total understanding highly unlikely in the near future.

Together this leads to what I have termed hypognostic ethical purposes. It means that all models concerning ethics, of future and past events are understood to be incomplete. I have chosen to create this neologism, hypognostic, because it refers to a total incompleteness of knowledge beyond the agnostic. The latter term is limited by its transcendent implications. I choose the "hypo "prefix to indicate incomplete knowability rather than a totally unknowable state of affairs. It differentiates between, but includes, the necessary uncertainties concerning our models of this P-law consilient universe and the unknowable transcendental.

Our hopes for the achievement of a personal goal and its means assumes a cause-effect relationship between the present and an imagined future. The fulfillment of these hopes is to be achieved by personal future interventions on reality in concert with possible outside forces. One further limit in determining wise goals, and our actions to achieve them, is what will be discussed under the term 'Free Will' (See §§ 13.11,13.12).

An imagined future without any proposed probable effective actions is the basis of emotional hope. Such hope is not a goal, but is, at worst, a laissez-faire cause-effect forecast. At its best, this proposed future outcome counts on causes beyond those initiated by our own free will. The proposed attainment of such an outcome is partly based on the imagined actions of others and nature. In religions, the others include gods and spirits working from a transcendent realm beyond our ken. Emotional hope is, at base, a statement of personal helplessness. This can then lead to feelings called hopelessness and despair, especially when the limits of probability approach zero. Both of these negative emotions are subjectively much less desirable than their positive partners, rational hope and purposeful action. Rational hope is based on uncontrolled, but possible, perhaps likely, future events, and is revisited in the Epilogue.

11.6 Aristotle and Goals

The thoughts of Aristotle have been a common starting point and basis of these kinds of discussions in the post-enlightenment philosophical world. One of his conclusions was that happiness – Eudaimonia — was a major goal of life.

" *Eudaimonia (Greek: εὐδαιμονία [eu̯dai̯moníaː]), sometimes anglicized as eudaemonia or eudemonia /juːdɪˈmoʊniə/, is a Greek word commonly translated as happiness or welfare; however, "human flourishing" has been proposed as a more accurate translation.[1] Etymologically, it consists of the words "eu" ("good") and "daimōn" ("spirit"). It is a central concept in Aristotelian ethics and political philosophy, along with the terms "aretē", most often translated as "virtue" or "excellence", and "phronesis", often translated as "practical or ethical wisdom".[2] In Aristotle's works, eudaimonia was (based on older Greek tradition) used as the term for the highest human good, and so it is the aim of practical philosophy, including ethics and political philosophy, to consider (and also experience) what it really is, and how it can be achieved."* (Wikipedia entry, 9/22/15, since expanded)

This rather long quote is included to illustrate several of the points I have made previously. First, there seems to be several possible and desirable goals in any one life. Secondly, the definition, never mind the translation, of each of these goals is necessarily variable depending on culture and language. It will also serve as a point of reference in reviewing the limits of evaluations, including ethical ones.

11.7 Happiness and Welfare

Happiness and welfare, are not synonyms – they are not even the same kind of state-of-being. Happiness — as in 'joy' — is a temporary state of mind that is felt more strongly than what is meant by a feeling of welfare or contentment. The former is a response to a recent, current or anticipated relatively short-lived event. How long can any mind be in a happy state? Minutes? Hours? Certainly not more than a few days. It is an emotion which, if more prolonged and of lesser intensity, is called a mood. Contentment, a feeling of welfare, is a state-of-being that is judged by the overview and then the summation of all of ones needs. This includes whether these requisites have been currently met, or are expected to be met on a more chronic basis. The needs include the physical needs for life, and the needs for mental harmony – the feeling "that all is well". However, this feeling of wellness, just as happiness or any emotion, is a temporary state of mind. Therefore, neither can be considered an achievable permanent goal. Achieving these aims of happiness or welfare, toward a more basic goal, drives much of our day-to-day activities. If this is indeed so, what could such a goal be?

11.8 Values from Mental Harmony

The 2013 book "The Righteous Mind" by J. Haight discusses the underpinnings of different moralities as exemplified by political (liberal, conservative, libertarian) leanings. He speaks of reason vs intuition as the two main determiners of the kind of morals one has. I would substitute the C-L and I-E systems, and agree that their relative strength and dominance affects the development of an internally consilient value system. Furthermore, the C-L system is usually used, by the typical adult, as a way to rationalize the initial, stronger I-E reactions. Less often is the case where the C-L system is used to evaluate, with available facts, that original response, and supersede it.

Broadly speaking there are three types of approaches to minimize mental disharmony concerning ethical dilemmas. The most common is that rationality is not called upon and one accepts "my gut feeling" as the arbitrator of morals and ethics. The second, less harmonious case is the hypognostic attitude, which accepts that absolute answers to moral questions are rarely possible, especially if they are at all complex. It is more often the cause of qualms when reviewing, after the fact, one's actions. At the other extreme, rarely found, is the person who tries to totally exclude emotional reactions without acknowledging the evolved fit between unthinking reaction and environment. This will lead to a judgmental attitude as strong as seen in the first case.

Unfortunately, the price of such certitude often leads to not only strong judgements, but strong, even overwhelming, actions. In short, a righteous dictatorship of the "left" or "right". Both lead to suppression, violence and fear. The hypognostic, who tends to be less assured of their conclusions, is then attacked by both sides. S/he does not subscribe to the dictum that "might makes right". This leads to the ethical conundrum, what does a hypognostic do when faced by a power whose strength derives from unwavering certitude. This question will be addressed in Part IV, but the answer depends on melding the best of the C-L and I-E evaluations, and then acting with the same certitude. Unfortunately, this usually means picking up the pieces caused be the destruction fostered by the wars between the hard left and right. It also often leads to inaction, or at best, self-preservation rather than a societal peace.

Haidt (2013) lists six pairs of foundational values; whose mix generates the moral attitude of an individual and the ethic of a group. They are: Care/harm; Liberty/oppression; Fairness/cheating; Loyalty/betrayal; Authority/ Subversion; Sanctity/Degradation.

I will not deal with these pairs individually, but only point out that both of the strong systems can maximize harm, oppression, cheating, betrayal, subversion and degradation. To minimize mental disharmony the emotional reaction "chooses", without pondering consequences. If a "pure reason" response is claimed, it means that one has chosen to discount all non-material consequences, because the emotional impact on other beings is uncertain and not easily quantified. At best, the hypognostic approach considers the broadest range of foreseeable consequence before initiating irreversible action. Actually, irreversible actions should ideally be avoided, considering the lack of absolute certainty, but this is often not feasible.

The positive list is much more valued, and called upon, when treating with the effects of similar actions on oneself and one's 'tribe'. Haight discusses this flip from egotistical "chimp" to groupish "bee" – but this is at best a metaphor. Bees are communal by instinct alone; chimps have the addition of emotional evaluations. The latter may have some slight C-L input, but neither is a good prototype for what humanity could be at its best, a best that will never reach the perfection of full mental or social harmony. The hypognostically acknowledged inherent limits are more fully discussed in Part IV, but will not satisfy those that seek certainty.

11.9 A Life Worth Living

A stated goal has been "a life worth living". Ethics is inherent in this goal because of the inclusion of the word "worth". One claim, attributed to Socrates, is that the 'unexamined life is not worth living'. This statement is prospective – and notably contains a double negative. A goal can be deduced from this statement. It is that the constant positive purpose that should be followed is the examination, which infers the study and understanding, of life. This refers to human life in general and specifically one's own. The phrase does not necessarily indicate any further goal or wider application, and therefore can be taken as more than an aim. Could it be taken as an end in itself? Yes, it can be a goal because it is terminated, by having an absolute ending; the cessation (death) of the individual human organism. The goal, would be to maximize ones understanding of the universe and oneself. It is a goal which claims no desired consequences beyond the limits of one's own life. As such it is a purely egotistical goal without regard to any social consequences. It is the goal of absolute hermits who neither depend on nor are depended upon by others as they imagine life's possibilities. It is a goal for an isolated person in a personal Eden for themselves. As such, it cannot be shared by others, especially on a planet that is hardly as forgiving as an Eden for one. For there to be a universal Eden, it would need to be able to supply all the needs of billions of human beings and their environment. This is not an achievable goal, but perhaps it could be an ideal. The end that would be supported by this ideal would be an increase in the wellbeing of as many others as can be achieved by one's own understanding and contribution to this effort. The transmission of knowledge will be found to be a necessary component of shared understanding and undertaking if progress is to be made to that end. But the values associated with the biologic needs are only the baseline. As discussed, in § 15.6, the mental needs are just as important. These needs go beyond the biological needs of cell life, but are dependent on it. These 'lives' of the brain-mind, such as "love life," intellectual life" etc. will be found critical in the discussion of death in that section, and why it is avoided and feared.

11.10 The Perpetuation of Life

Another goal might be the perpetuation of life in the broadest sense. Why would one make that a goal? It certainly is compatible with the physical reactions, tropisms, reflexes and emotions of all living things. It presupposes the actions that assist in providing personal safety, maintenance and reproduction. These are the processes common to all Reactive Living Things (RLTs), as reviewed in § 9.1. To this end, one should only add that we strive not to interfere, more than necessary, in the complex ecology that has evolved as a result of these common processes. Such allowable activity can be derived from Schweitzer's discussions concerning his concept "Reverence of Life" (Schweitzer,1969) (See § 13.1). The goal of perpetuating life, has set limits, with variation. The limits are set by the laws of physics, chemistry and the other emergent conditions necessary for life to continue to exist. The one, seemingly absolute, limit is the proposed entropic end of our universe. Much before that would come the fiery end of our livable Earth by the expansion of our sun. This change will occur when the sun runs out of hydrogen to feed its fusion furnace. In this scenario, our descendants along with a limited part of our ecologic system, might, with a low probability, avoid this fate by finding a way to escape the solar system before that particular, or another, earlier, end of livability on this planet. This is assuming that the environment of our haven, compatible with the requirements of living cells, will not cease to exist from self-induced destruction before such natural events occur. Any of these potential ends of earth-based life seem very far away in terms of human, never mind bacterial, generations. But why champion life in any case? The only answer I can support is based on its characteristics, as outlined the section on

Ontology and expanded on in the subsection on the ontology of mind (See §§ 4.1, 9.5). The process of life is the most prolific emergent source of variety in the kinds of possible being that we know of. That begs the question why 'being', that is existence, is a state to be desired. The positive answer to that question is based on our instincts and emotions, and not our intellect. And those instincts and emotions are central to the preconscious evaluations of the value of being, a subject of evaluative philosophy. Typical conscious cognitive evaluations add detail, but are not the basic mechanisms that underlie the most commonly held concepts of beauty, evil, and morals; nor in the choosing being as a goal. Neither a solely rational evaluation of these concepts nor a solely reflexive evaluation, although possible, would fit human needs.

There is a major difference between the questions posed by psychology and sociology; "what is typical and what is aberrant", versus philosophy; "what is helpful and what is harmful" in terms of discussing goals. Awareness of the concept of being is only possible to MLTs. If the question of being is not important, then neither is conscious life, without which the question could not exist. It is the possibility, of some being able to conceive this basic question, that makes the aim of preserving life in general worth pursuing. This non-egotistical philosophy is at its best when it includes information garnered by the social sciences, and they in turn are most useful when applied to a development of possible life goals.

11.11 How to Evaluate Harm and Destruction

One of the ways to approach the evaluation of harm is in the temporal dimension. Is the harm short-lived (minutes, weeks, years) or long-term (centuries, millennia or longer)? The longer a harm persists implicitly involves the relative (vs total) irreversibility of its effects. Short-term harm, may however be amendable to repair and limitation. It may also be necessary to achieve a greater aim or goal.

Destruction is the irreversible outcome of change. Harm involves the evaluation of that change. Change, by definition, destroys what was — replaced by something with usually increased entropy or by producing something with less entropy at the appropriate level of structure. This is also inevitable. For our purposes we will limit ourselves to the changes involving living things, including humans, their societies, and the larger environment. A few examples will illustrate this point.

All life is composed of individual cells, whose life demands change. Before the advent of chlorophyll early forms depended on energy contained in their watery environment along with the materials dissolved therein. Chlorophyll allowed harnessing the energy from some of the rays of the sun. This led to an explosion of life forms beyond its simpler beginnings. The debris of this multitude of cells, once they died, became a better source of material allowing evolution to favor newer and more complex forms. In essence the new materials formed a new niche. It should not be surprising that all life forms are related if one looks back 1 billion years. Initially, it is not inconceivable that different DNA structures competed, but the losers would be gone without a trace. Subsequently, the continental drift which tore apart Pangaea, the mass extinctions due to great environmental changes such as meteors and massive volcanic eruptions, created new niches again and again. Irreversible change of habitat is a central driving force of evolution.

Most living animals must destroy other living things to go on to propagate themselves in the proverbial food chain. A few depend on the disassembly of life caused by others. It is true that at the base of the food chain are the energy of the sun and in a few cases local, earthly, hotspots. But then the loss of the living structures follows, either by physical nonliving forces or by other life forms needing energy and material sources. Beyond those basic facts, ethics must consider the parable of Albert Schweitzer of the farmer, his cows, his field and the single flower on the way home. (See § 13.1) It is not destruction per se that causes ethical problems, but rather the reasons for its occurrence.

In social relationships a similar case can be made. Giving up multiple weak relationships for fewer (hopefully) stronger ones for the benefits of marriage, children and/or increased cooperation with others involves change, including partial or even total loss or neglect of previous relationships.

Understandably the complexities of life, humanity and societies makes an absolute certainty in our efforts an impossibility. We are, and always will be hypognostic, with limited understanding. But the best efforts must be made if we are to pursue our goals.

Chapter XII
The two Evaluative Systems and Aesthetics

12.1 The Instinctual-Emotive (IE) and the Conscious-Logical (CL) Systems: A Contrast

The discussion of the concept of aesthetics is often left to the end of any book of general philosophy. Here however, it will be discussed earlier because it is a commonality we may have with some other, though less neurologically complex, living beings. The reason for this view is that the C-L aesthetic valuation of a sensation is only an intellectual endeavor aimed at discovering why the human I-E system evaluates various stimuli on an aesthetic scale. It therefore presents a less complex and more basic analysis upon which the later discussions of human values that lead to ethics may be based. In the discussion of model making in ethics we will see that both the I-E and C-L systems are involved, making that task more complex. To review the I-E versus C-L systems, see Table II in § 9.11.

The ability to create reasonable models of reality by the use of logic is a totally conscious cognitive ability. This is in contrast with the aesthetic sense, which is created at the preconscious levels and derives from the I-E system only. Only subsequently does one further modify the initial feelings, and justify them, by using the conscious system. As with all types of evaluation, aesthetics ones are readjusted over a lifetime of experience.

Logic is useful because of the consistent and related P-laws that describe and limit the organization and relationships of the universe. Perhaps surprisingly, the same laws also underlie what one finds, for example, aesthetically beautiful. In both cases these abilities arise out of the emergent laws as found in the brain-mind complex. Both of these evaluative systems use the abstractions from the various stimuli, both internal and external, and derive value by the reprocessing of that information — but by different pathways. Feelings of danger and safety seem to underlie part of the development of the innate, instinctual reactions that eventually led to our development of an aesthetic sense that is unrelated to survival.

The efficacy of the C-L system rests on the quantum character of matter and energy, and the consistency of cause-effect relationships, allowing and reinforcing the creation of logic circuits in the cognitive evolution of our brain-mind. Logic circuits are rule-bound by the structure of the nervous system at both a cellular and molecular level. These structures inform both the possibilities and probabilities of model creation as well as the responses initiated by the preconscious adaptive mechanisms. Also, aesthetics via the emotions, provide one of the fuels for what we call will (as in "Free will", § 13.11,12.12).

By considering how these two systems influence one another I will develop, in Part IV, a better understanding of how they interact in creating the other evaluative functions of the brain-mind and then lead to action.

12.2 The logic of Aesthetics: what does beauty have to do with wisdom?
The judgment that something is beautiful, ugly, or aesthetically nondescript seems superficially to have little to do with philosophy's quest toward wisdom. This is because the connection is indirect. As noted in the beginning of this book, to be wise, is to have as full and deep an understanding as possible of a situation, especially one that is fraught. But as the saying goes: "Beauty is only skin deep"; even if one is speaking only of an apple rather than a queen. Apples and a queen feature this misapprehension in the Grimm's version of the fairytale of "Snow White". The evil stepmother-queen is, initially, the most beautiful woman in the land, as confirmed by a magic mirror. The poisoned apple she later offers to Snow White is also beautiful, and thereby tempting, to behold. This story also features dwarves, which, while not ugly, are atypical specimens — and in fairytales, perhaps not quite human. But — they are good.

At the end, it is only the prince, who is both handsome and good; although in the original Grimm version not averse to exact revenge on the evil queen by having her dance to her death in red-hot iron shoes.

Another fairytale that points out the potential and real superficiality of appearance is "Beauty and the Beast". In this case, the heroine's appearance and character are in synchrony, but the malformed beast, to all appearances, ugly, and even monstrous, is at his core, a good, fair and kind person.

One of the ways to be wise, is not to allow appearances to be the main, never mind the only, input for the evaluation of a person, thing, or situation. But neither can this form of evaluation be neglected.

12.3 Aesthetics-- The Relative Importance of an Emotional and Instinctive Evaluation

Beauty and ugliness are rarely, if ever, judged solely by conscious, cerebral circuits. If one could be totally, non-emotionally cerebral, like the character, Mr. Spock, of Star Trek, we would perhaps speak of well-formed and deformed, or safe versus harmful. I believe a good case can be made that a sense of beauty is felt when the same stimulus fosters a subconscious judgement of desirability, safety and appropriateness — and ugliness is the emotional response when the subconscious evaluation determines there is danger, repulsiveness or potential harm in what we observe. Fortunately, we have access to both systems and we can sometimes "see" beyond the level of the skin.

For the more complex evaluations there are good evolutionary reasons for the development of the two evaluating systems. The instincts and emotions react nearly instantaneously to sensory input, whereas the slower cognitive evaluations are more likely to create a model that is more detailed and accurate in its representation of reality. The conscious system is, however, used only to justify or modify the aesthetic reaction. That is, it judges whether the current reaction is justified by past experience and proven models.

The attempts to fully consciously codify which patterns of sensation underlie the feelings of beauty, ugliness or other aesthetic evaluations, via logic only, is therefore in vain.

12.4 Evaluations Beyond the Aesthetic

In general, when given the evaluative time, and the two systems agree, the likelihood is that such consilient models will better fit reality. When the two systems come to discordant conclusions, the cognitive response is often more correct, though not always. This is because some stimuli arising from the environment, and processed by the instinctive-emotive pathway, can selectively be given the wrong weight due to habit and memories of unrepresentative (of the current situation) previous encounters. Cognition's relative slowness allows for a wider variety of input from memory and a recursive check of which of several models better fit the abstracted stimuli. Also, memory can be checked as to whether, on previous occasions, the preconscious I-E system had led one astray. But as noted before, time is not on our side in many fraught situations.

12.5 Instinct and Emotion Are Often Right

Instincts and emotions are complex reflexes, molded by evolutionary forces that allow one to quickly respond to the environment in appropriate ways. A few examples concerning the main external and internal senses will illustrate this interpretation as outlined on in Table II in § 9.11.

Nervous systems, that are more complex than the relatively simple motor reflexes, allow for finer distinctions and integrations of external stimuli. These higher levels may include auto-stimulating from long term memory. Without here re-enumerating the levels of organization, the question becomes: at which levels do the processes that are called instinct and emotion arise? This will be important in any discussion of inter-species ethics. Part of these investigations rest on the parallelisms in these functions between individual H. sapiens, earlier evolving hominid forms and still extant model forming species.

Instincts, the preprogrammed nervous system responses to complex environmental stimuli, have developed because they presumably led to an increased viability of an organism, as well as an increase in its reproductive success. This is a property of many multicellular, multi-organ and nervous system containing species. Its emergence depends on a life form having all three levels of development. It also depends on having highly integrated afferent systems leading to coordinated and prolonged motor outputs.

Emotions can be seen as instincts that have, in addition, hormonal system responses which are initiated, reflexively, by the nervous system. This additional response prolongs and modifies the basic instinctual neuronal responses via non-neuronally derived chemical stimulation or repression. These chemicals in turn affect the NS as described in "Behave" (2018). This implies that all species which have a complex nervous system and an integrated hormonal system can potentially be said to have emotions. The consequent responses based on such brain-mind output differentiates feeling (internal mental phenomena) and emotional responses (externally evident phenomena). This distinction is important when studying these phenomena.

It should be noted that motor responses that imitate emotions, but that do not mirror the actual feelings at that time, is a form of acting. It is often an important part of effective lying and misdirection. Of psychologic interest is that the conscious, though false expression of an emotion, can actually support the parallel feeling via poorly understood feedback mechanisms.

12.6 The Senses

The order of the evolutionary appearance and development of the senses can serve as a scaffolding for the next sections. The earliest senses are the chemical senses, smell and taste; both are based on the more primitive chemical tropisms. Proprioception, the sense of directional contact was also developed very early. The stimulus is provided by pressure causing a deformation of cellular structures. This is very basic, if simple life forms are a guide. Vision, based on a sensitivity to wavelengths found in sunlight and hearing, based on subtle physical pressure on specialized sense organs arose later in evolution.

12.7 Olfaction, taste and the tactile senses

Of the two chemical senses, smell is the one most closely associated with emotions. It is also the only caudal (head based) sense that does not send its information to a primary area in the cortex. Rather the information goes directly from the nasal receptors to the forebrain systems where it is processed. (See Appendix for an outline of brain anatomy) When speaking of smells, we are less likely to use terms such as beautiful and rarely ugly, but rather use words such as pleasant or revolting. The sense of smell, for survival, varies tremendously in importance from species to species. In humans it is most useful in determining the edibility of food, the presence of decay and death or certain dangers such as fire. Many animals also use smell for detecting predators and mates. This broader applicability makes it questionable whether the emotive adjectives can have cross species meaning, but must rather be used in a specific (pertaining to a species) way. The sense of smell is also involved in the subconscious effects of certain molecules called pheromones. That chemical system of communication is totally subconscious in its operation, triggering I-E systems such as sexual attraction (https://www.ncbi.nlm.nih.gov/pmc/articles/PMC3987372/. This is followed by upstream endocrine and motor actions in a complex cascade. Persons also seem to have identifiable odors, and these are also associated with the emotions and instinct. Their effect on the cognitive system is usually subliminal in nature. Whether such chemicals should be called pheromones, versus odors, when they are unconsciously perceived, I leave to those that investigate this area.

The above and the tactile senses serve as a source of environmental information and communication. The tongue is involved in taste, and along with the fingers has highly developed receptors for fine touch. These three senses are of lesser importance than vision and hearing in the evaluation of human interpersonal and social interaction, although flowery perfumes, a malodorous breath or a soft touch can certainly affect the interpretation of close physical encounters.

12.8 Vision

When we perceive beauty or ugliness, especially at the extremes, they command attention. Here nutritional concerns may have played an early part. In fruits, that are noticed for potential consumption, yellows and reds often indicate ripeness — but when these colors are monotone and glaring, they can indicate a poisonous quality. Especially in the former case one speaks of co-evolution, although it is also true in the latter. Green is the color of healthy plant life; the dull browns often indicating spoilage or death. Regular and symmetrical structure also indicates health; deformity indicates injury or disease.

The yellows and reds and pale violets of a sunset, compared to dull gray skies on the horizon (in the middle latitudes), foretell calm overnight weather when the prevailing winds are westerly. An expanse of green is indicative of adequate rain for plant growth, and therefore for animal survival. Darkness is a friend of the predator; light allows our eyes to scan for danger and opportunity. Dull colors often serve for camouflage. All of the above are major sources of information that may foster a sense of harmony, security and near-term survivability, or can forebode danger or personal harm and lead to feelings of insecurity, dread or fear. The appropriateness of the instinctual actions that follow these sensations determine the success or failure of the individual in the short term.

12.9 Hearing

Given only sounds that are generated by the inanimate world, sounds can be as faint as a gentle rain or as furious as a volcanic eruption. The animate world emits sounds in a smaller range of volume, but with much more variability in pitch, purity and rhythm. First, we will concern ourselves only with sounds emitted by nonhuman sources.

Sound, like vision, is dependent on energy transfer over distance. The inanimate sounds that are of interest to creatures, are often a sign of danger. The crack of thunder, the crackle of a forest fire, the crash of a falling tree causes an instant alertness or a fear response. Most other ambient noises are mostly ignored (not focused on) by the amygdala. But the sounds that other creatures make are of more specific import. A few examples must suffice from the myriad of possible stimuli. Those that may serve as a warning, causing defensive responses, are the growls of predators, the hiss of a snake, the drone of wasps, as well as the unexpected snap of a branch or rustle of stepped on dry leaves. All of these can automatically cause the protective responses of freeze, fight or flight. The first response is especially important in weaker and less mobile prey — although predators that depend on stealth can also show freeze behavior as part of their own instinctive behavior, albeit without the emotion, fear.

There are also sounds that serve as attractors or have a calming influence. Although not the exact opposite of the protective responses, they have the positive responses of attraction, restfulness, and the wish to abide in such surroundings. Birdsong throughout the world serves as a warning to competitors and as an attractor to the opposite sex. Much of this vocalization is musical; that is, it is composed of purer pitch than that from most other animals, and is repetitive, melodic and rhythmic. Warning sounds in birds are, on the other hand, usually coarser, more irregular and louder. Other species have developed the instinctual ability to assess some of these variations, and react accordingly. Birdsong indicates that the singers do not sense danger and, given their high perch, are often correct. Therefore, the sound of songbirds indicates safety to others as well. The sudden cessation of song by many birds at the same time, or the onset of cacophony accompanied by the noise of rapid flight may indicate danger. It is this evolutionary history of sound evaluation in a natural environmental context that contributes to the emotional force of human music. We speak not only of beautiful, but of soothing, energizing, depressing and agitating music because of these auditory to I-E system links.

All these pre-logical stimuli-emotion relationships have been sketched above because they underlie, interact with and may even supersede the evaluation of stimuli by the conscious brain-mind. They are also the basis of all aesthetic evaluations. Conscious cognition can then evaluate the stimuli-emotion relationships, leading to further understanding and the possible modification of response. It must also be noted that conscious evaluation can cause instinctual-emotive responses by calling up memories or fantasies that imitate primary external stimuli.

In short, beauty and the other aesthetic valuations are sensed when the stimuli initiate emotional attraction, feelings of harmony and safety, or are the source for the fulfillment of other emotional needs. Conversely, sensory input that causes disharmony or aversion, a sense of dread or danger, or portends a lack of needed resources for life is major cause of the perceptions of ugliness and its kin.

12.10 The beauty of logic and art

One hears and reads about mathematicians, scientists, and other specialists who speak of the beauty of an equation, an experiment or a newly formulated concept. This usually occurs when the subject is either complex, or one that addresses previously poorly understood relationships. The cause for this kind of an evoked feeling of beauty can further inform us about the more common use of the term beauty as used when referring to the perceived aesthetics of sensory input, as discussed above.

In the latter context the use of the word points to the idea that the aesthetic evaluations are the reaction to the coherent complexity as recognized patterns. Here the sense of beauty is stimulated in the emotional system by the sense of accomplishment and understanding as the conscious mind realizes the import of the discovery.

As it is with ideas so it is with the arts. When it comes to the visual and auditory arts it follows that not all such 'art' is beautiful, nor is everything that is beautiful 'art'. What is art? This highly controversial topic only involves philosophy insomuch as it is a prime example of the problems of definition in all fields of endeavor that deal with the unquantifiable. The literary arts are judged more similarly to those in the intellectual endeavor contexts. It arises from clear and coherently understood feelings arising from the patterned complexity of prose, poetry and drama. In trying to define art, it seems one needs to involve the sense of beauty and the other aesthetic feelings. It is a slippery topic because of our current inability to quantitate the pertinent sensory input nor its effect on the preconscious mind. This will remain so until we can identify and quantify the neural patterns that are consistently involved when, for example, a sense of beauty is felt and reported. Until then it is a task for neuro-psychology to search for consistent patterns, of cross-sensory parallels, and attempt a fine tuning that eliminates the definitional uncertainties and seeming interpretational paradoxes.

Part IV
Toward Wisdom: Application of Knowledge, Values and Purpose

Ethics: "Do the right thing: Judgement of ethical action comes in two major flavors. One is typified by the Kantian categorial imperative, the other by Benthamian situational and relative utilitarianism. Many variants of both exist, but all have foundered on the expectation that a set of ethical rules could be found that meets all situations with acceptable results. This inability to formulate acceptable absolute rules, over many centuries, is strong evidence that it is not possible. This problem is not commonly acknowledged because in everyday situations most actions that have an ethical dimension are fairly straight forward. Furthermore, they are supported by cultural norms, akin to etiquette in formal society. Nevertheless, thought is given to ethical conundrums, and philosophy or theology-based systems are sought to give some guidance in the more complex or fraught cases. However, at best they can only provide a rule-of-thumb guidance unless the specifics are exceptionally clear. The following three chapters discuss this age-old search for guidance. One of the conclusions is that absolutes, as in mathematical axioms and the laws of physics, are not possible in the realm of ethics. In fraught cases only a wise evaluation, based on an ethic consilient with the real world, will reach a balanced solution. Chapter XIII discusses general questions concerning the purview of ethics and some definition of the central terms; Chapter XIV concerns the emergent neurologic foundations of ethical thought; Chapter XV considers some applications. The Epilogue gives reasons for the hope, of slow progress, in the quest for better ethical guidelines.

Chapter XIII
The Purview of Ethics

13.1 Ethics – An Overview

How humans should best lead their lives, both for themselves and for the world at large, is the purview of ethics and religion. As noted previously, worldwide religions and eastern systems of conduct base their conclusions on the received ideas of founding figures, who claimed either a transcendental source for their special mystical insight, or are built on revered cultural traditions and myths. Later, we have a more secular idea, such as those of Cicero's (106-43 BC) SUMMUM BONUM. These historical claims either lie outside a realistic, hypognostic world view or are antecedent to modern concepts concerning the sources of human behavior. These views do not seek confirmable cause-effect relationships between the claimed transcendent sources of action which override the universe's laws of physics (P-Laws) and their relationships. This heritage based; transcendence infused systems are all based on pre-science world views. And new, emotion and confabulation-based systems are constantly created by human minds – despite confirmable data to the contrary. At best these beliefs are naïve – at worst they are the result of wignorance. Unique, brain-mind generated insights may at times be accidentally true - versus logically based. But when not confirmable by objective referral to this world, such ideas cannot be a reliable input to the wisdom which is needed, in this overpopulated and prejudicial world, to formulate the best ethical system possible, applicable and helpful. And, there is no objective evidence of any universal purpose, nor external teleology, functioning on earth. All we have are fallible rules and suggestions as supplied by our C-L capable brain-minds.

The best ethical system requires a complex emergent value system, which will inform which actions are supportive (helpful) or at odds with (harmful) to the welfare of individuals, communities, humanity and life as a whole. The concept of values was treated in Part III and is reviewed there. The resulting derivation of a consilient ethical system is based on consideration of the relationships between an ultimate goal, its associated purposes and the derived, ethically valued, actions or means.

13.2 Working Definitions for Ethics: Means, End (Goal) and Purpose

The brain-mind sources of ethics will be covered in Chapter XIV. Here we proceed with a main-stream, concise definition of ethics as found in Wikipedia in April 2018. (Since then, modified with reference number changes. Italics and underlining in the original). Then the meaning of the basic concepts will be examined.

> "Ethics or moral philosophy is a branch of philosophy that involves systematizing, defending, and recommending concepts of right and wrong conduct.[1] The term ethics derives from Ancient Greek ἠθικός (ethikos), from ἦθος (ethos), meaning 'habit, custom'. The branch of philosophy axiology comprises the sub-branches of ethics and aesthetics, each concerned with values.[2]
>
> Ethics seeks to resolve questions of human morality by defining concepts such as good and evil, right and wrong, virtue and vice, justice and crime. As a field of intellectual enquiry, moral philosophy also is related to the fields of moral psychology, descriptive ethics, and value theory.
>
> Three major areas of study within ethics recognized today are:[1]
>
> 1. Meta-ethics, concerning the theoretical meaning and reference of moral propositions, and how their truth values (if any) can be determined

2. Normative ethics, concerning the practical means of determining a moral course of action
3. Applied ethics, concerning what a person is obligated (or permitted) to do in a specific situation or a particular domain of action[1] ".

Using this definition as a general reference point, we can embark on a more detailed discussion. All the above concepts are the product of the human brain-mind. As such they have the characteristics of mental models, but without the surety of the much more data driven axioms of the P-Laws. This will be seen to be important in the discussions concerning the application of ethics in Chapter XV.

First, a more general definition is proposed that covers the main points of normative and applied ethics, that is, acceptable conduct in a society. This is difficult when it is realized that there as many norms as there are social groups. To arrive at a universal ethic (See § 10.5) one must assume that all humans feel they are part of the same social group, an ideal that is highly unlikely to be ever met. A fully shared invariant emotional and instinctive viewpoint concerning relationships with different others is psycho-biologically impossible. It is not, however, impossible to pursue cultural norms that have broad support concerning acceptable ethical relationships. This aim can then be the basis of a secular ideal against which any proposed ethical system can be measured. This effort has led me to the following proposed definition of a subset of ethics.

"Normative Ethics and Morals: Ethical rules, socially created, resulting from the effort to normalize the expected conscious control of instinctual-emotional reactions. These rules are meant to facilitate acceptable habitual responses to the conundrums of socially important interactions. Morals are individual norms that are a subset of an egotopia. These may or may not conform to the ethical guidelines that are a major part of the ideals of a realistic or utopian social contract."

Applied ethics, including justice and prescriptive law, will be covered in §

13.3 Different Intended Actions: Good (Benevolence) Versus Evil (Harm)

Much of the historical discussion of ethics has been based on the values of various world-views, often based on the Abrahamic western religions. Some like A. Schweitzer (1936) have taken a wider view by including classical Asian thought. Those discussions provide the commonly held concepts concerning the meaning and source of harmful evil behavior and its positive alternative, benevolence. Especially interesting for ethics is "The Kural" attributed to Valluvar. A realist's ethics must therefore include an explanation of what those terms might mean, without a transcendental world view, and must define them for pertinent rational discussions concerning the fraught administration of justice (Chapter XV).

First, as indicated by the heading, for evil to have any ethical meaning it must include the concept of intent. Evil is often equated with, although it is not the same, as harm. Harm is defined in this context as structural or content changes that alter and decrease the possibilities or probabilities of positive value. Harm may also arise from changes in needs or resources or both. In an inanimate world one would speak of destruction or disruption, but not harm. The converse in this realm would be the facilitation of complexity.

Some animals seem to demonstrate intent, but the inferred limited concept of the future, in the subhuman realm, is the functional limit of evil behavior. Which animals, if any, have the appropriate nervous systems that can be potentially evil is difficult to ascertain. Evil should only be proposed when the harm is done with aforethought and without regard to the universal needs of the physical, (and in higher forms also the mental), processes of life and indirectly for its preservation. Because life is a part of this world, and general material destruction and disruption does, or potentially will, harm living things, evil can be done when the intended harmful action only directly impacts inanimate things. However, earthquakes and other natural disasters could only be evil, in this view, if caused by a conscious powerful being. This makes defining evil in religions such a conundrum. What imagined being intends the harm? In a realist's world view, only humans seem to have the potential ability to foresee enough of the consequences of their actions and therefore be accused of evil intent and actions. Human originated harm is currently limited to the earth but soon that realm of action may extend to parts of our solar system as well. The same is true of its antonym, benevolence; intended helpfulness. The same logic would apply to similarly endowed creatures elsewhere in the universe.

Harm from evil action does not only mean the initial outcome. All the foreseen consequences must be considered. The more powerful a person is, the more harm s/he can do, and therefore greater care must be taken in their actions. This is in contrast to the Machiavellian view that the powerful 'princes' must be judged less harshly. (Mackie, 1977). These considerations are especially pertinent if the first direct consequences are thought to be helpful to the universal needs of life. Unintended consequences must also be ferreted out as much as possible. Also, a willful neglect to counter harmful action, or to refrain from preventing foreseen harm, is the evil of passivity. Finally, wignorance is especially pernicious because it can masquerade as unintended, ordinary ignorance.

Small children, and some adults with atypical brain-minds, do not have the ability to foresee beyond the immediate consequences and are logically, and usually, judged less harshly for their actions. The same is true of all other animals, including the other primates. At which point a child's or another brain-mind is able to cross the threshold to an adequate understanding of future consequences is not clear cut. But it can certainly be achieved, to some degree, by the time children are able to logically discuss what future outcomes may be. The interference of emotions, especially strong ones, can be considered in our judgement of an action when the I-E reaction is immediate to a situation and not enough time has passed to allow logic to assess the case. However, in a professional setting, if multiple outcomes are foreseeable, and somewhat likely from personal and group experience, then training over time is needed to create ethical habits. This would hopefully create either better habits or foster the ability to find less harmful interventions.

A major ingredient to many evil actions is the expected positive egotistical feelings (brain reward center stimulation) resulting from contemplating the proposed action. These mental rewards include the positive ones of self-esteem as well having power. Self-esteem is provided by achieving desired accomplishments, but also by devaluating the accomplishment of others. The latter evaluation is a major source of Schadenfreude, the joy of seeing harm come to others. The positive feeling associated with power give rise to more active harm. Spitefulness, is a form of revenge, and can be so strong that it includes harm to the self. Nevertheless, the expected emotional benefits for the self may override concerns concerning both kinds of harmful outcomes, and the evil action is taken. The expected benefits may also include material goods that satisfy a personal desire. These two conditions describe evil in its fullest.

If the ability (for multiple reasons) to foresee the harm is weak, or if it was temporally impossible to override the egotistical instinct, then the action, although harmful should not be classified as evil. Here we may speak of thoughtlessness, carelessness or ordinary ignorance. Again, wignorance, the willful turning of a blind eye to reality, when it contributes to harm, is a special case of evil behavior because it is by definition intentional and avoidable. If none of these apply, and yet harm is done, we may speak of accident. Because of these difficulties in teasing out such nuances, the presence and level of evil should be ascribed only after a thorough investigation that looks at all ascertainable facts. This will be necessary to obtain an approximation to full justice, the social response to all harmful actions, as discussed in Chapter XV.

Passive evil, a weaker partner of active evil is part of the concept behind the German word "Schadenfreude", the joy derived from seeing harm, or lack of success, come to others. When the harm caused is not under the control of the self, yet enjoyed, then this can easily become the precursor to personal evil action. This passive term indicates less interaction and may accompany willful passivity, that is when a personal action to ameliorate that harm could be forthcoming. The latter, an abdication of responsibility, could be called inactive evil, a much-neglected concept. It underlies what H. Arendt (1994) included in the meaning of the "banality of evil" (See § 13.12). Another passive but more personal partner of the preceding is when, retrospectively, harm from one's own actions is noted, and no remorse is felt nor countermeasures are taken. The latter is also a form of evil. It is treated much too lightly in our society because possible action versus actual action is much harder to assign to an agent. The results of various actions may well be the same, be just as good or harmful, but in order to determine the possibility of an evil intent, the above logic should be applied. As A. Schweitzer (1960) noted, we all cause harm, to each other as well the rest of the living and inanimate world. (A short version of his example is found in § 13.9). Often this is inevitable when we act to protect our own health and life. But he also felt that one should always feel sadness and remorse concerning this necessary harm. If not, the habit of self-forgiveness can soon allow real evil to be done.

13.4 Medical Decisions and Ethics

One area in which the need to separate harm from evil is the field of medicine. This short discussion will hopefully provide some clarity to this distinction. In western medicine the phrase "first, do no harm" is consistent with parts of the Hippocratic Oath, though not explicitly in it. This dictum should be evaluated in view of the above outline concerning the intention of helpfulness when necessary short-term harm is done. It is clear from the P-laws that all action causes change, and thereby the destruction, that is loss of, the previous condition. Ideally, the net consequences of all available actions (including the lack of action) must be weighed against each other. Does "Do no harm" mean taking only actions that are meant to facilitate the short-term consequences that foster commonly held helpful goals? The immediate harm of an injection, a surgical incision, or the side effects of chemotherapy for cancer cannot, should logically not, stop the initiation of such an action when it is expected to be beneficial in the final analysis. This can be seen as a utilitarian answer, but it is broader in its scope than the classical usage of the term. A medical intervention is primarily performed to help the individual patient in some way. But this individual may derive both positive value and negative consequences to themselves. This in turn may impact the community and potentially the world. It is this broad set of outcomes that makes absolute rules or ethical imperatives unworkable on a universal scale. The field of medical ethics must constantly deal with these issues. However, even here, religious values are often taken into consideration, values that can cause real worldly harm in the hope for a proposed, but highly uncertain, transcendent benefit. In any case, acknowledging our hypognostic state of knowledge furthers the view that final absolute answers are hard to come by, if not impossible to make. This is especially true of ethical conundrums where a long-term beneficial outcome, with some known short term harmful consequence, is proposed.

Finally, the lack of a commonly held end, by those that act and by those that are acted upon, makes it impossible to avoid differences of opinions on the best way forward. This is especially difficult in those controversial cases when scientific data is unavailable, contradictory and as always, incomplete. The following sections will deal with these issues and outline the limits of ethically allowable actions. When those complex ethical evaluations are discussed, the somewhat simpler evaluative attempts at defining aesthetic concepts such as beauty (in Part III) should be borne in mind.

Triage – an extreme case? This book was nearing completion in the years of the Covid-19 epidemic. Hospitals, medical personal and the general infrastructure of health care where often overwhelmed around the world. The concept of triage, born on the battlefields of the past, entered the awareness of the general population. Triage, choosing who is helped and in what way, and to what end, is an amalgam of the practical and the ethical. It pits the needs of the individual against the larger pertinent group. This is a major reason why only one ultimate end can offer any hope for a defendable position. But only if that end is agreed upon.

Without an overarching goal any decisions made would be arbitrary, short sighted or both. Such an untethered state brings to the fore the fraught endeavor of evaluating the meaning, the potential consequence, of individual lives whose fates are brought together by disaster. Only the goal of a thriving community in a thriving world can be a solid foundation to the decisions made. But, because of our hypognostic limits, the best medical decisions are made by those who humbly apply their wisdom, derived from the medical and psychologic sciences as well as a beneficent ethic. Only these decisions and their derived actions will stand the test of retrospection.

13.5 The Effect of Distance and Power

There is a general issue in ethics which is not often considered. That is the distance, in space, time, and psychological relationship, between the actor and the acted upon. Although the intensity of effects is not exactly the inverse square, as it is with light and gravity, it is an important factor in the ultimate consequences. The harm or helpfulness possible is also proportional to one's power over another or a situation. It follows that those with the most power should be especially aware of the consequences of their actions. There is more detail in Chapter XIV.

Distance and power also explain why it is harder to keep in mind, and extrapolate the effect on those that are thousands of miles away, those of future generations, and all those with whom we have no personal contact. This may be a minor issue if one's activities are only of local consequence and in the foreseeable future, or if one's power is limited to the typical range. It is also easier to judge. For all of us the closer the impact distance, and the greater the force, the more important are ethical considerations. The powerful must be held to an even higher standard of thoughtfulness and intent since the effects of their actions can and do affect situations much further away in space, time, and social connection. The ethics of the strong must therefore come under special scrutiny, and the importance of ethics in these cases should be seen as major factor in determining society's response to their actions.

13.6 Expanding on Definitions as They Pertain to Ethics

Goal, aim, purpose, and aspiration are related concepts. In these discussions, the word 'goal' indicates a desired ultimate future end result. If more than one goal is entertained, this will almost certainly generate unresolvable conflicts in trying to create a consistent set of ethical actions. This is unavoidable in large diverse social groups such as nations. I therefore propose the need for an ongoing search for only one universal goal. This ideal will never be reached, but it highlights the harm caused by argument compared to a cooperative collective search. Such a search will serve as a useful purpose by promoting helpfulness over harm.

Aims and purposes are subservient to the ultimate goal, and although often used synonymously, they are separated herein for clarity. Aims are all the proposed intermediate steps to achieve a given, often implicit or even unrecognized, goal. They are often linked in a temporal order, with a later aim being dependent on the successful achievement of the earlier. When the specific goal is noted, aims are the proposed means toward that goal. They are less central in ethics than the goal because alternatives pathways, often mutually exclusive, may be available toward any specific goal.

Purpose points to conscious awareness of the relationship of aims to the goal. One may therefore speak of the purpose of an aim, namely the achievement of a recognized goal. The goal has no purpose because it is the final summation of an ultimate idealized future.

An aspiration is defined as a more nebulous imagined outcome than the above, but one with a stronger emotional component than desire. Although emotion provides the drive, the will, to an action, its inclusion does not allow a purely logical defense of the aim or goal. One's aims and aspirations are open to context, and are less easily defended, than the goal, because they are situational to the path taken. They also have differing plausibility of success depending on the proposed pathway to the goal.

An ideal goal, as used in this discussion, is an end point that stands alone, without regard to aims needed, nor influenced by emotion, but is logically materially possible. It comes closest to the imperative of Kant. If one proposes two or more goals, all but one (at most) would, on analysis, be seen as an aim. Only a thorough parsing of the competing "goals" will help to choose one of them as a possible universal candidate.

There are some further differences between these concepts. Calling some intermediate proposed outcome an aim indicates that it is part of a group of related actions in pursuit of a common goal. Purpose is a more active noun in that it implies the will to action and implies an emotional component. Ethical and functional evaluations and considerations can be most specific for those aims that are limited to the immediate future. Some aims may, however, be proposed in service to an imagined future that does not consider physical possibility or ethical norms. If found to be in service of the former we approach full fantasy. If truly free of the later we are in the realm of "pure" science and technology; that is calculations of possible future states, their possibility and their intended and foreseen functional capabilities without considering value. In science the purpose derives from the emotional need for better models of the truth. Ethics can then determine if the aims of a scientific endeavor are consilient with the ultimate goal. Ethics is used to evaluate the desirability, a value judgement, of competing pathways containing the proposed aims. Desirability is itself a product of emotional and logical evaluation. The lack of ethical consideration in science, and especially in technology, was highlighted historically by the development of the atom bomb. "Pure" science is dangerous. If specific scientific aims are to be a benevolent service for our world, these aims must be ethically judged.

All typically proposed candidates for a common human goal and their means should involve the use of the best available models of a proposed future. It must always be with the understanding that the hypognostic provenance of the 'facts' will limit our certainty. However, commonly ethical evaluations are not derived by just the use of the C-L system, but also the emotional, preconscious evaluations of the proposed possible action. In any case, the norms of an ethical system should be checked for internal consilience before being applied. But as noted earlier, the unconscious I-E evaluation may be at odds with the C-L result. This common disharmony may make a purely logical ethical action impossible for any particular actor. Even so, specific personal goals have been given the force of an axiom. This makes discussions unfruitful.

13.7 The Place of Symbiosis – An Example of a Muddy Definition

A post-delivery human being is not just a clonal collection of cells in very specific relationships to each other, but also contains, and needs other unrelated organisms to function and thrive.

The recent findings concerning the effect of the human microbiome on our lives need to be accounted for. Its effects on our nervous system activity are especially important for philosophy. Although Deepak Chopra (2019), an alternative medicine advocate, calls on findings of non-DNA factors as evidence of another form of intelligence and an expansion of the human DNA, this is not a useful use of these terms. A form of intelligence that is not based on the ability of concept formation is a non sequitur. Adaptability is a better word in these cases.

Symbiosis is a form of biofeedback, often between cellular systems that have different life courses. It was coined in 1877 by Albert Frank to describe the relationship between algae and fungi in lichens (Sheldrake, 2020). Bacteria and animals are completely different life forms, both on a genetic heritage basis, and by the lack of a nervous system in the former. To interpret "intelligence" and "human DNA" in such an unlimited way can only cause confusion and poor mental modeling in a reader. What can be emphasized is that there are many environmental influences that can, and do, affect our brain-mind that are not mediated by our classical senses. Teleological mechanisms should not be invoked for tropisms and biochemically based symbiosis. But such interactions show the importance of the greater ecosystem in the modern discussions of philosophical topics. This awareness of the great influence of other life forms led Schweitzer (1960) to propose such concepts as a common life-affirmation, also referred to as the will-to-live, and led to his promulgation of life-devotion. Together they encourage the emotional and intellectually acceptable "Reverence for Life". "Reverence for Life" and life-devotion requires a human-like brain-mind. The will-to-live is, however, a nebulous concept that combines a mental property (emotion) with a functional one (instinct). One might then speak of a boat having a will-to-float, a teleological inference. Only time and further research will give us more information needed to develop better ethical models that take non-teleological symbiosis, in addition to human-human symbiosis into consideration.

Whether mitochondrial DNA, which is in our cells but not a part of the nucleus, should be considered 'human' is less clear. It is believed that this DNA was obtained eons ago by the incorporation of early bacteria into the forbears of all plants and animals. In some cases, there is even evidence that bacterial pieces did eventually become part of the nuclear DNA of all eukaryocytes. But in any case, DNA is not a form of, nor has, intelligence. It is a molecular chain of nucleic acids (with supporting structures) forming a highly stable molecular pattern that composes the information which allows and underlies the structure and function of all living cells. Proposing a molecular level emergent property that conforms with intelligence is a case of an exceedingly loose definition. This confuses rather than clarifies the relationship between lower-level causes (molecular chemistry) and higher-level effects (neurological properties). This error arises by ignoring all the intermediate levels of structural complexity that lead to the emergent properties at the level of the brain-mind. It may be helpful to review §§ 4.5 and 4.6 at this point. Symbiosis is another example of complexity, emergent from the interaction of higher classes of matter, that need not involve teleology.

13.8 Primacy of the End Over Means in Ethics

That *"The ends do not justify the means"* is often taken as a basis in a discussion of ethics. This section discusses the alternative. *"The means do not justify the end"*, is just as fraught.

Ethics deal with the interactions between a person and their family, friends, larger circles of acquaintances, strangers, animals and the rest of nature of this earth. The inherent enormous complexity of possible interactions belies the belief that absolute rules of action(means) are possible. Unresolvable conflicts between all these possible relationships inevitably arise. I therefore discuss a system of an ethical end rather than the primacy of means, as championed by Kant (See § 14.8), and examine why and how this informs allowable ethical actions in any particular circumstance. The primacy of an ethical end, versus means is obviously a shift away from Kant's categorical imperative, though not as much from the utilitarian ideas of Jeremy Bentham. The primary difference between my approach from that of the latter is that it is a search for a more inclusive absolute, singular, end goal. This is a complex endeavor, but it recognizes the inevitability of having to choose between possible conflicting proposed and absolute ends as well as the myriad aims and means available in moving toward one final goal.

From a transcendent perspective the ends are the wills of the gods, demons or spirits, as ascribed to them by prophets and seers, both old and new. These ends differ widely and cannot be judged by the P-laws because these other-world views assume axioms that are by definition in addition to and beyond the P-laws. This imposes the impossible task of choosing between different, contradictory fantastical models on the basis of which of the many self-described authorities one has been exposed to. Such beliefs, which are models of ultimate ethical authority, are primarily shaped by the cultural factors that influence personal emotional reactions and needs, that is the I-E system. This may be workable for self-imposed moral percepts, but cannot serve the needs of a complex society composed of many subgroups, both local or worldwide.

The question then becomes whether the same can be said of any secular system of ethics that is developed by the efforts of human brain-minds in the search for a consilient set of reality-based ethics. If the emergence of ethics is seen as an evolutionary consequence of the development of the human brain-mind in a social setting, then the attempt will be seen to have the following strengths and weaknesses.

A major strength of such attempts is based on the fact that there may be as much certainty as the internal consilience as the data, gleaned by science and common experience, allows. A perceived weakness is that it will never be final, and that it can only be as strong as the logical connections between the data allow. But, by always striving for consilient models which explain the actions that result from the output of human brain-minds, these attempts can then lead to an ethical structure that may be usefully debated. Besides never being complete, such formulations will lead to what has been called, and is often derided as situational ethics. I propose that this property is actually a strength, because it can lead, with effort and wisdom, to a defendable answer given to ethical questions. It is however recognized that the answer may not, in a particular situation, always be consilient with the emotionally desired outcomes to the question of "What should one do?" especially when one considers the many possible valued means toward a common end. The "should" then refers to the answer that is derived from the considered, and then determined, value of these conflicting means. Initially this value will differ the most on the personal scales of I-E evaluations. Only continued thoughtful discussion will slowly lead to more agreed-to answers. As in science, the answers will be incomplete at some level. Also, because the I-E system is the ultimate source of the will, the logical means may never be implemented. (See also §§ 13.11, 13.12 on Free Will).

13.9 The Central Place of "Should" in Ethics

In ethical evaluations of an action, should is "used to indicate obligation, duty, or correctness, typically when criticizing someone's actions". (Oxford Living Dictionary (now Lexico.com)). For Kant the priority of duty was central. But his source of the laws of duty is outside our physical universe – he still assumed a transcendent sphere of authority.

In realist philosophy, duty is seen as a utilitarian means toward an end, even if not formally ascribed to that school of thought. However, in a transcendent theological framework the source of duty is an often-unacknowledged utilitarian evaluation. Here the utility lies in the human wish to please a god or spirit, or conform to the assumed desires of these imagined transcendent beings. Obviously if these beings were not ascribed the ability to affect our lives (the biological or the assumed life-hereafter) there would be no point in conforming to their supposed rules and wishes. So even a Kantian ethic has a utility which, as in realism, lies in conforming our actions to some inferred consequences!

Albert Schweitzer was a modern example of a person who tried to meld the two world views in his actions. Two instances from his writings (Free, 1988) must suffice here as an illustration. The first comes from an agrarian perspective, the second from a religiously felt duty to serve mankind.

Schweitzer tells a story of the farmer who carelessly and unthinkingly knocks of the heads of wildflowers on his way home after cutting hay in his fields. He finds the former wrong and harmful, but the other, even more destructive of flowers, justifiable. His logic is based on the necessity of most fauna, a form of life which he greatly values along with all others, to consume other, once living things, in order to maintain and propagate themselves. The hay is for cattle, whose milk and meat sustains many humans. The destroyed flowers on the wayside were harmed for no commensurate, or justifiable, reason.

The personal sacrifices he made to open and maintain a hospital in the wilds of African Gabon, at Lamberéné, were based on his belief that his immense talents were a gift of God, and he needed to use them for the good of the world. He furthermore decried, and found hard to understand the unwillingness of the natives to help others outside their tribes, even in simple helpful tasks such as transporting a sick person on a stretcher (Schweitzer, 1985). He felt one had a religion-based duty to use one's abilities for the common good, specifically in this world.

In any case, it is the present and a preferred future, as imagined for ourselves, as well as for other beings and things we care about, that drives ethical evaluation. These preferences (organized ordinally), underlie the evaluation of what should be, or should have been, an appropriate and justifiable action.

The underlying values and mental processes that inform these preferences, need to be discussed from an objective point of view that contains as little emotional or cultural prejudice as possible. Preferences are not physical laws and are therefore not invariant between individuals and their cultures. By invoking "should" one must accept that such conclusions are inherently situational and relative.

13.10 Humane Pragmatism

One result of the above discussion is the acknowledgment that, inherently, all culturally based ethical norms lead to an unavoidable "relativity'" problem. At one extreme, a specific goal and its aims are judged, without the use of logic, by one's own sense of morality (inherited both biologically and culturally). At the other extreme, judgement is only affected by the logical value of the means (aims) as related to some unitary absolute goal. This extreme lies outside any consideration beyond that ultimate goal. This is the "the end justifies the means" approach. The Kantian imperative approach is that "the means justify the end". Either absolute can lead to psychologically abhorrent results when pursued to a reductio ad absurdum conclusion. In contrast a humane ethics is a method that considers the combined typical needs and abilities (the mechanics of the means) of cognitive beings. (See § 14.6). It is not relative to a particular culture or individual human history. It could also be called "humane pragmatism". (It seems to be loosely related to Dewey's views (Tiles (1988), Shook, 2000), but I have arrived at it by broad general reading (Legg,2020; Menand,2001; Misak,2020) and following the possibilities and limits outlined in Parts I – III in this book. It was not developed as a point-by-point commentary on Dewey's voluminous work). The adjective, humane, is borrowed from its normal usage – "marked by compassion, sympathy, or consideration for humans or animals" (Merriam-webster.com) but is here specified to also have its source in both the I-E and C-L systems of the brain-mind. It also largely overlaps with Schweitzerian ethics. It is not to be confused with a humanism that places mankind alone at the apex or center of all possible or potential beings. A humane ethic is a consciously derived ideal, a description of devised rules that allows the attainment of the commonly held and necessary physical needs of human life along with the supporting interpersonal and other environmental relationships. As such, a humane pragmatism is based, in part, on the logical limits of probable outcomes of actions that maximize the survivability of the species. Furthermore, a humane ethical end would be the goal against which all proposed or actual actions are weighed. Since we, as living organisms, depend on the interrelationships between us and our total environment, living and physical, our aims must also include the stability and appropriateness of the future environment as a necessary constituent. The same could be said of personal morals.

A personal moral goal is limited in that it is a vision based on an individual's unique set of conceived aims and goal, an egotopia. Likewise, when a full set of common personal purposes is shared with others it can be the basis of a utopia of a specific group. The ideals of both egotopias and utopias are by definition not a humane ethical system because they exclude many other groups. They are usually also self-defeating due to internal inconsistencies. Ideals, even humane ones, have the further attribute that they are either unachievable due to the limitations inherent in reality, or at the very least, are highly (nearly infinitely) improbable. Some future personal desired event that is not limited in this way can be a concrete aim, which then can be judged by referral to a near universal pragmatic ethic.

The goal of the survival of the human species and its supporting environment, must follow from aims that include consideration of mankind's dealings with, and its impact on, the broadest environment. This requires interspecies ethics, because one of the aims must be the fostering of the viability of earthly life-in-general. This critical to future human survivability beyond a minimal biological existence. To be human is also to have a human psychology, one that appreciates and needs beauty, contentment, predictability, and the curiosity fueled by the variability in the environment. Humanity is at this time, by far, the greatest source of action that can satisfy these needs. Without similar levels of these needs it would be the life of a sub-primate mammal, (never mind a reflexive reptile), experienced minute to minute, meal to meal. It would steal away those characteristics which makes us human. The note of the lack of these mental needs in an otherwise physiologic human becomes pertinent in the ethics of imposed death (§ 15.6).

As an aside, it should be mentioned that the very far future may include the necessity of leaving this planet. This may be well before the incineration of Earth by an expanding Sun. But for our purposes we can rationally only consider the near (~ the next one to one hundred years?) future.

13.11 On the Concept of Free Will: Statistical Randomness or Cause-Effect Relationship

On the macroscopic plane in which we live and act, I don't believe random (uncaused) chance, is supportable as a mechanism. When the word, random, is used in this context it only points to a lack of sufficient information to fully foretell the future or explain the past. Even at the subatomic, particle physics, scale such uncaused change would clash with the axiom of zero-sum energy/matter conservation. Only in the equations concerning the hypothesized sub-particles, as in string theory, does uncaused chance seem to be a probability. This possibility is lost at the next higher particle physics level due to the process of emergence. Because of this we only consider the possibilities that emerge at the atomic particle level and above.

Viewed from the present, all past occurrences had, by definition, a chance of 1.0. All the events that didn't happen, although they were logically and potentially possible, had in retrospect a chance of zero. The size of a calculated statistical chance is based on limited, well-defined outcomes, which are usually mutually exclusive, and based on a finite set of data. Real life is not so simple although the calculations are usually based on a simplification of the totality of that reality.

To illustrate with a question: "What is the chance that this typical and fair flipped coin will land "heads" on this table?" Mathematically the simple answer is usually given as 0.5, but in reality, it is less. For example, it may fall on the floor, or someone catches it and carries it away, or in a rare case it may even bounce and come to rest on its edge. So, the true answer lies somewhere between 0.5 and zero. This is how it is in life. The rules of nature are fixed, but our knowledge of all the pertinent forces acting on the situation is usually significantly incomplete. Fortunately, most of the time things go as predicted, due to the most common probabilities, or else one wouldn't dare to initiate any action. One knows "enough", but the comfort level of enough is a psychological variable, different for each person and each circumstance. "Free will" at this level should not be taken to include uncaused randomness, but rather a choice from highly uncertain cause-effect outcomes that are in the realm of the possible.

The conscious execution of a "freely willed" action therefore depends not only on choosing between different imagined futures, but also on a full understanding of possibilities and their relative probabilities. The main necessary assumptions in defining all outcomes of an action deemed to be freely willed are as follows. It is assumed that a sentient being can a) extrapolate the present into the future based on current conditions and knowledge of cause and effect, and b) can mentally compare the outcomes, and then c) chooses the one that seems the most desirable. This is followed by actions that affect the near future facts(causes) in the hope to affect the resulting effects, in such a way, that it comes closer to one's goal, aims and aspirations. It cannot mean that we are free (unaffected) by our C-L and I-E capabilities.

Some questions then arise. Are we really able to choose based on a pure logic that is perfect? That is, can we even follow an unswerving computational algorithm? Or, are our biological wired tendencies and capabilities, such as temperament and intellect, formed and reformed by a largely uncontrolled (by us) environment? Are we really free to do anything other than what the brain-mind allows in its interactions with the external environment? Do we have enough information to say?

Often one can, in retrospect, figure out why one did such and such, especially, if it went against what we now think we "should" have done. This is an intellectual analysis of the past flow of actions and its presumed causes. This would include recalling our emotional state and the thought patterns at the time. By comparison, others are especially prone to judge our past actions from a more information rich present. This is what the American "Monday morning quarterback" does. But realistically, one goes through a lot of life using habit, "flying on automatic pilot", and one can hardly speak of free will in those circumstances. This is akin to how we partially seem to "will" the course of action when we are in a semi-alert dream state. In discussing a sports game, "he could have" is only a statement of pre-action, inferred possibilities. The actual, post hoc, probability and the possibility of the "could have" was of course zero and a different outcome is now a wishful illusion. The "will" while dreaming is seen by most western adult persons as an autonomous, unchosen

activity of our brain, whether otherworldly beings are invoked or not. If we become conscious of will the dream is, at least temporarily, over
.

13.12 The Extent of Free Will and the Concept of Evil
We now proceed to further define these two component concepts of free will independently and then consider how the former modifies the latter.

The word "free" is usually taken to mean without constraint. But an absolute freedom, with no constraints, is not possible for us under any conditions. As noted in physics, it is highly doubtful that there exists any object in the universe that does not interact with some force from another object in some way. We therefore need to speak of "levels of freedom" in philosophy. This is not the same as the "degrees of freedom" in statistics. Instead of concentrating, as in statistics, on the number of independent factors necessary for a particular outcome or state, the philosophical levels of freedom indicate the actual number of forces that contribute to a particular outcome. The lower the level, the more contributing factors there are to a result. This also means that a low level of freedom means less certainty as to causes. But the causing forces, while additive in a vector-like way, are not equivalent in importance. The influencing factors may include just one dominant one, but there are usually also a near-infinite number of lesser, even infinitesimal, forces at work. This conclusion is inherent in the fact that although our emergent macro-world is what we experience, it is built on, and emerges from, the nano world of interacting particles and energies. However, for making predictions at our temporal and physical level of perceivable interaction, those factors that drive about 95% of the outcome will usually suffice. Unfortunately, when it comes to social interactions, the obvious factors may be a significantly smaller part of the total pertinent input. All voluntary actions and relationships are therefore the result of causes that are, at a minimum, limited by the physics of the possible and probable. Even then, some major factors may be unknown, misjudged in their importance or even denied by a wignorant evaluator. We are left with a freedom that is limited to choosing a "best bet" among the possibilities we have imagined. Freedom is relative, and literal full freedom is a falsehood.

What of the word "will" in this context? Not as used in "I will do it" to indicate a promised future action. Rather as in "I will be purposefully doing it" to indicate that the action arises out of my conscious intellect, with emotional support. Besides presupposing consciousness, a purposeful will includes the concepts of imagination and a sense of the future. However, it is the emotional force, generated by the I-E portion of the brain mind, that powers ones will to affect the self and environment, in the attempt to bring about the imagined future.

Without the imagination and a sense of future we are left with reflexes and habits. Also, without emotions we would indeed be only automatons, our consciousness only calculating and observing results and reacting to those results in a recursive fashion for initiating the next set of actions. We implicitly are aware that an emotional force is available and necessary to our "will to action" when we hate and love, are afraid or suspicious. When we are totally at peace and content we tend to, indeed, continue on an automatic or at best a laisse faire basis. When we are apathetic or "do not care", we do not have the will, the drive, to change or seek change or redress. To say one cares, states that an emotion-based evaluation has been made.

So, what can we know about free will? Knowledge, as defined in Part II, is composed of mental models. These models are based on current external stimuli, and/or memories created from previous encounters with the environment. Together they provide the data that is processed by the I-E and C-L systems. Furthermore, new models are made by an imaginative process that favors the integration of all these inputs into a consilient whole. Putting these concepts together we come to the conclusion that, at maximum, will is the force that enables us to carry out purposeful actions, intended to bring to fruition a chosen imagined future state, a proposed model, which we care about. The actual outcomes of these actions will be limited by the P-laws of nature as well as the emergent properties of life, often unknown and unexpected. To will some action is an attempt to bend the present toward an emotionally preferred, hopefully logical, imagined future.

But what of the levels of freedom that restrain a total freedom of will? I believe these constraints are much larger than allowed for by most commentators on this subject. The first constraint that arises in the act of "foretelling" possible futures is one's imagination. One cannot have a full knowledge of all possible choices. To have full free will, in this sense, one would need to be omniscient. Since no person has that quality, we only have varying amounts of options, as prepared by our imagination, consciously open to us. This limited subset of all physically possible actions confines our actual choice of potential directed action. The options we can be aware of will depend on our personal history of direct and taught experience (and the correctness of the taught material), as well as memory, mental state (including personality, phobias, emotions), time available before an action needs to be made, and of course general mental capabilities.

This long and incomplete list could be expanded greatly in its detail, but it follows that the actions activated by the will are not free, that is, without any constraint. These limitations are not acknowledged in secular and religious law or even in our own, too often, misjudgments of the actions of others. That no physical force or threat of force was used to coerce an action, that would otherwise not have been taken, is a pale freedom.

The second problem arises from the central place of emotional/instinctive mental reactions. These preconscious evaluations determine the strength of will that determines the effort made to affect the future. That one may be "weak-willed" or "strong-willed" is not questioned by most observers. It is the source of that weakness or strength that is often misperceived. The level of will provided by the instincts and initial emotional evaluations is, by definition, outside any freedom of choice.

Does this mean that our actions are predetermined? Does this mean there is no evil? The first question is not answerable at this time in view of our incomplete understanding of the influences that cause us to ideate or imagine a course of action. It can, however, be said that the factors that go into which ideas comes to the fore in our consciousness are based on a very large number of current and historical influences. This number is so large, that for all but the simplest (mostly habitual) thought patterns the prediction of the reaction to a complex problem of conduct is indeterminable, even to oneself. The more complex the history and the less creative the imagination, the less likely it will be that the consequences of an action will be correctly foretold. Furthermore, the mental paths that lead to an imagined future, even if foreordained, has to be taken in subjective time, and each fork in the road has to be reevaluated, as memory adds more information to the original dilemma. A simple example: If a person "freely" associates on a starting word, multiple times in tandem, creating a long (say 100), derivative list of words, the content and order of the list will not be the same again and again – although probably similar, especially in the first several iterations. This introduces the idea of "chance", often misnamed "random", associations at a neurological level. As noted in § 3.12, the analog-to-digital conversions, and back, in the activity of the brain-mind is not totally predictable. This does not rule out ordered, probabilistic outcomes at the macro level that mirrors the same limited level of predictable certainty of the nano-level brain-mind activity.

The second question demands a definition of evil. As Bunge (2003) argues, evil is an adjective, not a noun. Evil actions and thoughts are meant to harm for the sake of, or in willful disregard of, harm. Evil exists - but it is the relational adjective describing the will to harm. The emotional basis of the conscious decision to cause harm or help, as discussed under free will above, will be important in a discussion of justice.

How does this square with an adage such as "Everyone is the best that they can be" - or the related, more common, and active "One does the best one can"? Secular evil deals with human intent and aforethought. It differs from theological evil only in the source of harm, namely a transcendental source. This is even encoded in insurance policies that may not cover certain "Acts of God". Secular evil is based on socially defined ethical standards of acceptable trade-offs of different harmful and helpful outcomes, rather than revealed or imagined ideals of behavior. There is no "problem of evil" in rational philosophy, only in theology. Hannah Arendt (1994), speaks of the "banality of (human) evil" when referring to the Nazi horrors, also speaks of the transcendent evil of the great 1755 Lisbon Earthquake. But she does not clearly differentiate them as belonging to different views of reality.

Philosophy, especially political philosophy, has to deal with how to apply the concept of secular evil to social interactions and how to deal with them in an ethical way. This demands much data concerning how and why humans interact with, and react to, their social and general environment. The Greek philosophers had a modern (in the sense of typical personal experience) understanding of human nature as it applies to ethics. However, this branch of philosophy, if it is to go beyond that personal view, will only be to be furthered by the inclusion of the voluminous data being built up by the use of the scientific method in all areas of inquiry. The alternative models, based on the belief of transcendent spheres of influence on the real world, often create a different set of derived ideals of behavior. The latter models of behavior often serve as a source for unachievable, albeit often secularly desirable, life affirming ideals. The following chapters on Ethics and Justice will expand on these concepts. The meaning of statements concerning the "can" of behavior depends on the definition of that verb. Can, in a secular behavior context, indicates a possibility; it does not indicate an expected certainty of meeting a goal or its associated aims.

Chapter XIV

The Source of Human Ethics

14.1 Needs, Values and Satisfiers in Ethics – the Complexities

The biological needs of life, and the resulting values of their satisfiers, was discussed in § 10.1. The already high complexities of those bio-needs pale before the concepts that make up the psycho- social needs which ethics and morals also address. In both cases the relationships can be summarized by the formula $V \sim \frac{N}{S}$. (Chapter 10.1).

Here we review and expand on those concepts. A Need ("N") always refers to a specific condition. The Satisfier ("S") and Value ("V") in the relationship refer to that specific need. But to speak of that specific need in isolation is useful only to clarify the relationship between the three related concepts. But then:

1* There are many needs, call them N1, N2, N3......NX, standard symbols for related variables. Each N has one (or a set of) Satisfiers and the consequent Value(s) of the Satisfier(s) or the Value of the set. As the urgency of a Need increases (↑) or decreases (↓), with the Satisfier(s) held constant, the Value of the Satisfiers will necessarily go up (↑) or down (↓).

2* Just as there are various needs there are S1, S2, ... SX, and various sets of them, available at different times. We will call the sets ST1, ST2 ...... STX, where the T subscript refers to the sum of all the individual Satisfiers in each temporal set. It is even more complex because the Availability(A) of each (S) can vary from situation to situation, time to time. But for simplicities sake we can note that the VS1 ($\downarrow$) whenever the S1 ($\uparrow$), and so forth. And of course, with an ($\uparrow$) AS1 there will be a ($\downarrow$) VS1 as well.

3* Therefore, in the simplest case of one need and one satisfier, we have:
V1 ~ urgency N1 **/ S1** where "~" indicates correlation.

4* In the case of ethics, we rarely deal with cardinally measurable Needs (as we can when talking of a chemical equation) and therefore have, at best, ordinal results for the Value of the Satisfiers. This is conceptually similar to the inputs of a chemical reaction: specific chemicals, amounts, temperature and time allowed for the reaction process. So, at best, "what should be done" will vary with circumstance of relative Needs and Satisfiers even if we have an ordinal Value outcome in mind.

5* As another complication we note that some Satisfiers relate to different needs in different ways. For example, a moist food (say an apple) can be the satisfier for both energy needs and water needs as felt with hunger and thirst. But the two needs cannot be compared to each other in a cardinal way. The relationship between the two needs is ordinal, for example: "I am hungrier than thirsty".

In summary, the urgency of biological needs can be approximately calculated and the availability of satisfiers can be enumerated and determined fairly accurately. On this basis the value of the satisfiers can be functionally determined. In contrast the psychological needs of (say) contentment versus anxiety are thoroughly subjective. However, the neurologic correlates are now better understood and can be measured as relative intensities by the crude (for this purpose) use of fMRI and other non-invasive methods. For example, does a virtual reality wolf cause more anxiety than a virtual snake in a particular person? Or, at the time of the test is a banana more desired than the aforementioned apple? But such determinations are preliminary and primitive compared to the assessments of the biological needs of an organism. The psychosocial needs for a functional community are even less ascertainable. This short discussion is meant to emphasize our hypognostic state, especially in ethics. The certainty implied be the word "should" in ethics is therefore not supported in reality.

14.2 An Ethics Based on Needs and Therefore Values

Our morals exist to protect that which is of value to us. And, as discussed above the value of a satisfier is related to the ratio of need and the availability of the satisfier.

I will posit that the lowest level of any need, is zero - on a sterile planet. Without the process of life, the concept of value has no place in an ethical sense. It is true that we speak of the need for carbon to make a diamond, but I wish to differentiate the word 'need' from 'necessity' in the cause-effect sense. I wish to limit the word 'need', when discussing morals and ethics, to a perceived necessity by an advanced nervous system. And, an advanced NS requires all the emergent properties of life. Ethics are therefore concerned with the needs of a living world and the safeguarding of its satisfiers.

So, what are these needs and some of their satisfiers?

1* For the general process of life we need the appropriate levels of water, elemental building blocks, energy sources and viable ambient temperatures. The upper and lower limits of these needs will, and do, vary between organisms, but have limits concerning these constituents. Even the famously resilient tardigrades, evolved to their high level of survivability from more fragile life forms, have limits to their survivability. In sum, from the level of niches to the world as a whole, there are always minimal needs for various life forms.

2* We have ever greater number of variabilities, associated with the increased levels of complexity, from unicellular species through to those organisms that evolved over time. Some thrive in arid niches; others need a permanent aquatic environment and others need or tolerate both. The same can be said for ambient temperature ranges and energy sources. In the latter case we have the sun for plants, oceanic thermal vents for some simple organisms, and other living organisms (current or antecedent) for most other species. Together, acquisition of these energy needs can be construed as a form of parasitism. But evolution added another level, that of symbiosis. Mankind uses both. An example of the latter is the bacteria that make vitamin K (which our own biochemistry does not make) in the specialized niche of our gut. Without this vitamin bleeding would be a constant problem.

3* Mankind is the most creative of all environmental modifiers, creating more and more suitable niches out of previously hostile and poor environments. This has been, and will continue to be at the expense of other organisms and the environment that supports them – and us. Therefore, purely material degradation must also be an area of ethical concern.

4* As with all social species, we have developed symbiotic relationships within our group; another level of needed interdependence. The preservation of the satisfiers of the needs of such closely integrated groups is the second major area of ethical concern. Related is the minimization of parasitic (versus symbiotic) behavior.

5* The needed ethical evaluations are based on our C-L or I-E mental processes. By definition it cannot be the I-E system alone, otherwise one would have to posit that all animals with a NS would have a form of ethical evaluation. This surely stretches the purview of ethics too far. An ethical system needs at least one or both of two capabilities of a C-L system: the ability to imagine different outcomes from a starting point (retro- or prospectively) and the concept of cause and effect that connects the past from its future. Some other social species do seem to have this ability, to a limited extent, as shown by observation and testing. A recent excellent exposition is found in De Waal (2019). But clearly, our I-E system is also a factor. It provides the automatic "gut feeling" to specific interactions: real, virtual or imaginary. It also provides the power of the will to create change in proposed or recurring situations. Different evaluations of the situations, by the two systems, are the major cause of our ethical conundrums.

As noted above, the complexities of competing needs increased in the eons of evolution from single cellular organisms to competing multiorgan species, then further to those in symbiotic relationships, and finally to those that can imagine the future. Therefore, the need for ethical evaluations also increases depending on the scope of needs, availabilities and variety of satisfiers as well as the level of competition and cooperation in the effort to maximize a valued outcome. There is also one more variable in the intensity of the ethical problems. This is the physical and temporal distance between an initial action and its direct or indirect consequences.

This long list of complexities makes the search for Kantian absolutes futile, except perhaps in the simplest of cases. We must however not be dissuaded from identifying guidelines that allow individuals and groups to fulfill their potential as creative beings in a competitive environment.

14.3 The Needs and the Human NS Methods of Satisfying Them

As noted previously, values are based, in part, on needs. Physiologically, all living beings share most needs and have appropriate capabilities for meeting them. Any living cell or organism needs a source of water, nutrition, and a thermal environment compatible with and supportive of the functions that allow it to grow, thrive and reproduce. The advent, 1.89 million years ago, of Homo erectus added, to the impressive list of reflexes, instincts and structural forms of 'lower' animals, the emergent abilities of the growing Prefrontal Cortex (PFC). This sporadic but cumulative PFC enlargement contributed new properties, possibilities and probabilities to the hominid brain-mind over the extended time during which brain size and complexity increased in various lines of descent. These mental properties, abilities and needs are studied and named by many different academic subgroups of neuroscientists, psychologists and sociologists. Sociology in this context can be seen as psychologically based interactions between groups. Psychology, and neuroscience in general, studies all the mental processes of separate subsystems of individual brain-minds, as well as the interactions between them, as they respond to their external and internal environment. To which academic field the individual interactions or group interactions belongs is somewhat arbitrary and most often driven by the historic, academic divisions of labor. These divisions, however, are not important in how the results of their studies are applied to philosophy's search for understanding and wisdom.

The mechanisms necessary to meet both the minimal physiological and psychological needs of a person are therefore more complex and nuanced, when compared to the processes of brain-minds of creatures with less memory and cognitive power than those available to a typical modern human adult.

Considering all this, ethics is seen as a group's conceptual set of guidelines of behavior as developed, over time and in different places, from the ongoing codification of the actions allowable to a human being in meeting personal needs in a social setting. It is an account of how various actions impact the needs of other life forms, especially other humans. Furthermore, since the earth's physical environment supplies all the resources to meet the physiologic needs, the effect of actions on the non-living environment is also pertinent. (That would include that portion of the sun's energy that impacts the earth.) Anything of value is therefore related to and proportional to one or several of these needs. The instinctually perceived value depends on the relative essentiality of those needs at that point in time. But, our conscious awareness of the future does demand the expansion of ethical concerns beyond immediate concerns and needs. Furthermore, as needs and resources change, the value based specific ethical formulations will, perforce, also change.

The more complex primate societies became, the less adequate became the mechanisms based on the Preconscious Adaptive Mechanisms (PAMs). The complex set of psycho-physiologic reactions that we call emotions guided the next level of organisms by integrating and interpreting the inputs from the environment, both internal and external, in a preconscious process that involves major NS subdivisions. Emotions are part neuronal reflex and part hormonal (non-neuronal, inter-cellular, and chemically mediated) response. Emotions and their subtler and longer lived, cousins, the moods, allow for responses that are consistent over a period of time. This increase in temporal capacity allows for a more coordinated and measured response beyond the classic 'freeze, fight or flight' paradigm. These three actions are the primary automatic responses to danger. Other PAM responses, such as approach, search, acquire, defend, and investigate are reflex and instinct based actions whose performance also requires reactive brain-mind stability over time. The relatively more recent moods and emotions improved this stability, compared to the PAMs' alone and allowed more nuance to the possible reactions.

14.4 Balancing Needs in a Variable Environment

There are many types and levels of needs which must be addressed to make wise decisions concerning ethics, justice and even day-to-day concerns. Over the course of this book, we have dealt with the personal physiologic and psychological needs, the latter being driven by instinct and emotion. As also noted, when conscious decisions are made, the models of the PFC's logical circuits are often in conflict with the concurrently formulated subconscious models of the PAMs and emotions. This will always be a problem for decision making.

We also have needs, as individual groups and as a species, which depend on brain-mind activity beyond the PAMS and I-E systems in order to survive, prosper and reproduce. For the purely personal needs we must cope with, and often even compete with, the nonhuman environment. With the social needs, personal conflicts with others of our group, and especially others outside our group, are added complexities that may come to the fore. Finally, and broadly speaking, come the common needs and resulting conflicts within and between groups. Here we need to deal with the problem of how to interact with individual outliers and hostile groups. We must however remember that the latter also have needs regarding their stability. That there is a very large number of pertinent interactions that can affect the outcome of any decision or action, wise or unwise, cannot be denied.

For all but the most recent eras of humanity's presence on this world the number of interactions with the environment needing to be considered was relatively small compared to the complexities of today. As a member of a local tribe any person's needs were met by personal effort, as well as with cooperation with a small number of fellow tribal members. At this level of existence, the actions impacting the nonhuman environment were (and in some cases still are) usually local in cause and partially reversible in effect.

Occasionally the large, near or distant, perturbations of volcanoes, earthquakes, tsunamis, major storms, epidemics etc. interrupted the usual flow of problems and associated decision-making (McMichael, 2017). The basic causes of these major disruptions were unknown or ascribed to mysterious forces, spirits or gods. If occurring less often than every 3 to 4 generations the successful previous responses were usually lost from memory as survivors died out after the calamities had been dealt with and overcome. Being infrequent, these disasters usually do not affect the gene pool, in an adaptive way, because the causes of death were indiscriminate relative to the genetic attributes of a population.

The more or less frequent interactions with other tribal groups could be divided into two main categories. Those that were proximate and dealt with daily to monthly and those that were further away, less frequently dealt with, or new. These interactions could be friendly (useful), tangential, avoided, or unfriendly (harmful) to the point of mutual destruction. At the local level, depending on the size of the group, the same divisions occur, but here the number of variables become more manageable for processing by the PFC. Because of this, the PAM reactions can be more easily modified or suppressed by conscious evaluation. In the case of habits, they will usually be well-adjusted to the day-to-day stresses of living.

Ethical wisdom can therefore be discussed at the local communal, area-intertribal and "modern-world" level. The latter would include, at its lower borders, the last 5000 to 6000 years of human existence. Wisdom can further be divided into those areas of judgment that deal with the human-human versus the human- nonhuman environment. The latter is important, but has been relatively neglected by philosophy until recently, even as the detrimental effects of the Anthropocene era have become more and more evident. This broad area of necessary wisdom will be discussed in the last section, "A Rational Hope".

In dealing with human interactions alone we can form four subtopics: person-to-person, person to intra-tribal, personal to extra-tribal and intertribal. In the modern world it has become more and more necessary to include the latter two in philosophical discussions.

Ethics, justice and their codified derivative, law, can be analyzed on a similar basis. In all cases the interrelationships and overlaps between the levels must be kept in mind. Furthermore, the developed judgments must be consilient with each other so that artificial paradoxes are not created.

14.5 Typical Shared Human Purposes Leading to Ethics.

An instinctive brain is not enough to create an ethic. If it were, we would have to argue that all mammals-and even some earlier animal life forms would have ethical ideation. That would make the application of this concept much broader than it is classically or currently conceived. However, human ethics are informed, in part, by our instinctive brain whose processes underlie our initial success as a genus. The conscious ideational evaluation of a situation (real or imagined, with or without external stimulus) served to further the more recent success in the continued propagation of our forebears. Whether our super-success is desirable is highly debatable (Harari, 2014). The topic of ethics highlights the fact that the evaluations of the instinctive and emotional (I-E) brain-mind are often in conflict with those of the conscious and logical (C-L) systems. As noted in Section 10.1 these two evaluative systems may be in harmony — or not. A positive correlation leads to a higher sense of rightness, but if they differ significantly or conflict, then the lack of correlation becomes a source of distress and confusion. It may also lead to some seemingly paradoxical conclusions in the area of ethics.

Early Homo genus ideation was based on a very limited set of information that included the sense of self, the community of tribe or clan, the existence of other nearby tribes, and a small section of the natural world. The uncertainty caused by the increasing abilities of the C-L ideational evaluation system was inevitable. This is because the human mind will try to adjust this ideational evaluation to the evaluations of the complex reflexes. It tries to form a consilient evaluation. This process is important to achieve perceived harmony, clarity and certainty in the planning of actions. An important mechanism in this regard is confabulation, which is the creation of pseudo-facts, that is mental data unsupported by normal sensory input and treated as models of reality, without being aware of their source in fantasy. This did not lead to too many difficulties at the early stages of our evolution since, in those eras, it mainly involved balancing the needs of the self and the shared needs of the community. When the needs of the biological cell (and its organ and system conglomerations), personal mental harmony and the perceived needs of the local community coincide, then the combined evaluations and instincts that are stimulated lead to unquestioned action. The confluence of the various needs was then best served by cooperative action. However, when the needs of these three systems are in conflict, and the resolution is not part of an instinctual set of behaviors, then standards supplied by a communal ethical system will be helpful, or even necessary, to avoid destruction of the community by the "strongest" individuals. "Strongest" in this context would include those with the highest physical strength, the most forceful will and the best manipulators of information and belief. These agreed to ethical standards, or rules of thumb, whether oral or written, were and are necessary for the long-term harmony with, and success of the self within, the community. Otherwise, chaos leading to the harm of either or both will eventually result over historic time. Over time, populations increased in size and number of sub-communities to which an individual belonged, or was excluded from, increased. This gave rise to the ethical conundrums that have bedeviled mankind since the advent of the larger physical communities such as the city-state.

In such an increasingly complex social environment an ethical evaluation of a situation often implies, rather than creates, an awareness of an underlying conscious goal. This inferred goal is, in fact, most often a subservient aim. This is because, while it is a product of an ethical calculus, it is in reality only a cognitive interpretation concerning underlying sets of instincts and emotions. The instincts, inherent to the species, are the givens that drive most of the actualities of human interrelation behavior, followed closely by successful habits. And, the intensity of the behavior is greatly influenced by the emotions. The harmonious possibilities, between the instinctually versus the logically derived attempts at ethical justification, become increasingly limited as more and more data becomes available to the C-L model forming process. Furthermore, the axioms of science, the result of the distillation of data to their most basic constituents and properties, led to a consistent and consilient worldview. But they also limit the potential harmony between the two evaluative systems because the instinctual and emotional systems are not confined by reality testing. The overriding goal and its ethically justified means of attainment, should be arrived at by an extrapolation of the possibilities inherent in living beings, and by an amalgamation of instinct, emotion and conscious intellect. But too often we find that personal and socially shared aims are only the result of basic instinct, reinforced by complex emotions, without considering those limits that our conscious intellect tells us are inherent in the P-laws of reality. What all these limits are is an important task in determining what is ethical. The task is not merely finding justifications of purposes that are based on purely animal instincts and drives.

14.6 Limits on a Common Secular Humane Ethics
Some human ethics are not humane!

We will start with some statements of givens that should pass the examination of non-transcendent worldviews. These are based on logical conclusions derived from scientifically gathered information over the last several hundred years.

Humans are, biologically, animals that have evolved from earlier forms of primates. Psychologically they stand out on the evolutionary tree. The earliest forms of the genus Homo seem to go back more than 1 million years, a time span that allows ample time for step-by-step evolutionary changes to arrive at the form and properties of the modern human brain-mind. Current data supports that the morphologically new species, H. sapiens, developed between 200,000 and 300,000 years ago. Language, a product of further mental and physical vocal development, is estimated be a factor after about 50,000 years ago. Also over time, and with spatial separation, certain sub-types based on superficial phenotypic variations developed. These surface variations led to the current vague concept of race. There is no clear data that any of these visible differences correlate or affect the processes performed by the common, typical brain-mind, except that generated by the influence from fellow human feedback. The influence of parental, peer and culture interactions have been shown to affect the post-natal growth and development of individual brain-minds, and by extension to defined sub-groups. Whether statistical variations have arisen in the mental talents of biologic subgroups is highly controversial.

However, none that affect the biologic basis of ethical development have been found by neuroscience. This gives rise to the following conclusions:

1* Our brain-mind has evolved from pre-sapiens, in a gross anatomy sense, by a marked growth of the PFC and the attendant growth and modification of the pathways that connect these newer structures with those of the midbrain and hindbrain. The latter two areas of the nervous system are structures (see Appendix) whose functions kept our very early forebears alive and successful in propagating themselves. This was accomplished by the actions mediated by of reflexes, instincts, and emotions inherited and modified from even earlier forms of mammalian, and ultimately their pre-mammalian life patterns.

2* Our most recent, non-Homo genus, forebears (however defined) needed to be a social species. This statement follows from the convergence of several major facts. The bodies of the larger primates are not useful for a solitary predatory existence. They did not have the speed of many four-legged mammals to protect themselves against other predators. Neither were they small enough to evade death by hiding in burrows or other natural shelters found in their habitat. Finally, they were not large enough to deter common feline and canine predators, in the manner of elephants, rhinos, etc. This left the possibility that the success of the group would largely underwrite the success of the individual. Without this social dimension the propagation of an individual genome would have been highly fraught

3* The larger head size, that was concomitant with the increasing size of the brain, led to the birth of infants in an earlier stage of development in order to accommodate the structural limits of the birth canal which remained proportional to the body size of the adult female. Infancy and dependency therefore had to last for an ever-greater number of years to allow the brain to grow until it reached its adult size and abilities. Of note is that even after gross physical size had been reached, certain functions of the PFC do not reach full maturity until early adulthood; males later than females.

4* The level of possible ideation achieved by the most recent nonhuman forebears and the clearly human forms of even 50,000 years ago, were reached gradually, step-by-step, not in large increments. There were many evolutionary dead ends of the Homo genus, as shown by fossil evidence. The changes from 50k to 5k years BCE are harder to elucidate. Major factors influencing the development of appropriate mental functions included the increasing size of clan groups, and the more frequent interactions with other humans that were strangers.

5* There has been an explosive increase in the human population starting with the advent of farming and animal husbandry about 5000 years ago. This increase has been exponential especially in the recent past. (See Wikipedia topic "Estimates of historical world population" for details.) Natural selection of the brain-mind I-E component has not been able to adequately adjust to the new social environment. This became especially problematic as daily interaction with many strangers became the norm, leading to increasing, interpersonal stresses. The C-L systems seemed, relatively, to have been able to keep up with this new volume of sensory input and information. This could be accomplished by new concepts being incorporated into local languages over time. As noted previously, however, not all languages include similar concepts. This can affect difference in response to the same stimuli. We subsume many of these differences under the term, culture. Most importantly, no individual can even begin to comprehend all the modern concepts concerning technology, physical science, social science, history, art, and so forth. This directly affects the development of personal morals and understanding, or lack thereof, leading to variations in culturally based ethics. But, the individual variation of I-E and C-L abilities to form concepts independently is greater than typical total abilities of any major social group.

6* Individuals share most of the history of genetic evolution (as expressed in their DNA). There is little difference in those genes that control the cellular functions shared with other multicellular life forms. More variations are found in the DNA and its expression that deals with macro-structures such as body types, soft tissue structure and the properties of the outer covering of each clonal collection. The genetic variability that underlies human talents and other raw abilities is less well defined, but the difference in the basic DNA pattern that underlies the instincts, emotions and typical intelligence of brain-minds seems to be minimal on a population level.

7* Much less shared is a common cultural history. The least shared is the temporal, experiential, individual environmental history which clearly impacts the development of each individual brain-mind, especially in the early developmental years.

8* Although the factors of each individual history can be referred to in the attempt to explain individual morals, we can only look usefully to the commonalities when examining and discussing the basis of human ethics. If a set of ethics for all of typical adult mankind is wanted, then only the shared histories can and should be taken into consideration. An acultural humane ethics, would come closest to a universal functional ideal. All the other local systems of ethics are, by definition, subsets confined to groups with relatively well-defined and shared common histories. These local ethics will by necessity vary in multiple ways from a humane ethics that is applicable to all times and circumstances. A common humane ethics, as discussed in the previous chapter, is broader in application, but more restrictive in details. Its will be discussed in Chapter XV.

14.7 The Neurologic Conflicts of Means and End

As discussed in § 13.6, purposes are cognitively and emotionally created models of desired future states for the self and the world. Means are the actions proposed and undertaken to achieve an ultimate end.

The main conflicts in creating one's morals, and by extension a system of general ethics, are based on what those means and end should be. And then, how the priorities between the means could be ordered as one compares the consequences and side effects. These conflicts arise when: a) more than one short term aim is considered (as one must do when responding to various conflicting needs) and b) the intermediate aims (imagined state of affairs, but logically leading to the primary goal), conflict. The second kind of problem also arises, even if there is just one end-goal chosen, because the intermediate, but necessary (to the goal) actions and their consequences may conflict with each other or the goal. How this can work out in practice will be discussed in the next chapter.

14.8 The Impossibility of an Absolute Ethics.

The best-known modern attempts at a final, controlling basis for ethics are the categorical imperatives of Immanuel Kant and the goal of 'The greatest happiness of the greatest number' popularized by Jeremy Bentham (https://plato.stanford.edu/entries/bentham/). The latter became the basis of utilitarianism, one of a large number of isms. This profusion of philosophical schools without a standard use of words for the same concept (or even worse, using the same word for different concepts) often leads to inevitably unresolvable contradictions. The lack of a clear declared definition of how a word is used by individual commentators only adds to the confusion. Dictionaries for some schools exist, such as Bunge's Philosophical Dictionary which "adopts a humanist and scientific standpoint". (Bunge 2003. Foreword). Others can concentrate on specific philosophers such as Kant because his writings do contain some definitions (Caygill (1995) or Spinoza (Runes (1951). Generally, textbooks and other comprehensive works list multiple meanings without trying to search out the most supportable in a consilient world view. I have tried not to fall into that methodology and hope to have had some success.

Kant divided his imperatives into the hypothetical and the categorical. The hypothetical is framed as an "if I want X then I must Y", and is interpreted as a moral for the self. It is about the "best" means, not ends. Whether X is desirable is not addressed. In contrast, as defined by Kant, the categorical imperative is a maxim that must be obeyed by all at all times under all circumstances and is meant to supply a universal maxim of action. In other words, it is proposed as a maxim that should have the force of a law of nature and is also about actions, not an end. The conflicts between the many absolute bio-psycho-social needs that lead to perceived necessary ends, is not addressed by this formulation. This categorical imperative formulation is one of the prime examples of an ethics that necessarily contains conflicts and paradoxes when more than one maxim is entertained.

Kant discussed the difference between what is right versus what is good by the use of "pure practical reason". He also noted that desire is the root of the will and that the will is a law unto itself. His conclusion is that the ethical law prescribes a duty, applicable to all. A well-known quote in this regard is "Act only according to the maxim whereby you can at the same time will that it should become a universal law without contradiction". Also, he prescribed the dictum "treat every person as an end, not as a means to another end." This seems to imply that ends, especially other human beings, have precedence over means. It does not tell us what the "end" of a person may be. This implied primacy of ends versus means in ethics is discussed more fully below. Furthermore, Kant was critical of moral intentions (feelings) versus rational thought. In many ways his conclusion is similar to the biblical Golden Rule which states "Do unto others as you would have them do unto you."

When taken as a central tenet, Jeremy Bentham's "greatest happiness...." has developed into a philosophy of utilitarianism. It has produced many logical results which are highly unpalatable. This is because by emphasizing a balance of happiness it does not address the depths of misery that are allowable to achieve this goal. Its other great weakness is that it excludes all entities that are not part of the human race. Finally, the parameters of this one end cannot be measured or even imagined, making for an absolute end that is nebulous and therefore gives no real guidance.

The "ethics" of the early Greeks was more about how to lead your own life. Its best-known formulation is that of Socrates. It is reported by Plato that just prior to his death, Socrates stated that "the unexamined life is not worth living". The import of this phrase was discussed in § (11.9). The contradictions between systems abound.

14.9 Ethics and Wisdom

It seems clear that cognitively-based purposes should be based on the best available information. Therefore, a prime aim of humanity should be an increase in and an improved accuracy of the information that reflects the workings of reality. A secondary aim is to decrease the loss of such information.

The key to these two aims is the concept of accuracy. Although in general one wants to increase the total amount of information, one also wants to set aside outdated or inaccurate information by replacing it with data that seems better grounded in reality. The same process also holds for mental models that are built up from this information by the creative process I have previously subsumed under the term imagination.

A useful comparison can be based on the DNA: ecosystem relationship. The totality of DNAs of an organism (and resulting RNA, proteins etc., with their functions) is the information. The resulting cellular ecosystem is the direct end result. The greater worldwide living ecosystem is in turn built up from this expansive cellular base. Similarly, the models built from sensory input and memory by the imagination can be seen as the intermediate building blocks of the sought for end result: Wisdom. Furthermore, just as a wide variety of functional and mutually supportive organisms based on a wide variety of DNA based structures makes a supra-cellular ecosystem workable and stable, so a wide variety of reality-based and consilient mental models may lead to a common wisdom. Application of wisdom can in turn lead to a stable and worthwhile personal life and its supportive community. It allows for the hope for a livable world now, and in the future. In all cases a lack of balance between all the levels of organization can only lead to long-term dissolution.

Death is the final personal organic dissolution and the extinction of the organism's clone, but the conclusions of wisdom may survive if passed on to others. A humane, but unbalanced, ethical system based on an inadequate application of wisdom, may well result in a premature induction of a sterile planet earth; the final dissolution of the living world

Chapter XV
Justice as Applied Ethics

15.1 Introduction
Wisdom, as guided by pursuing a rational consistent philosophy, is sterile without action. The solo actions of an individual, whether un- an- or ethical, are limited by their power over events. The ethical norms that govern the actions of groups, often at the behest of a leader, are even more influential because the combined power and resources are usually much greater. The justice system of a group implicitly enshrines the values and ethics of that group or its most influential subgroup in the complex societies of today.

15.2 The Pursuit of Justice
Whereas morals and ethics are mental models, it is the actions based on them that directly influence and create tangible relationships with the world. Morals, that is mental models of possible but unfulfilled action, are harmless- even if malevolent, and useless- even if wise. Ethics, being common to a group, are much more likely to lead to broad instrumental action.

There are some necessary components for an action to have an ethical component. Although these qualities have been discussed in previous sections it will be useful to list them here. The action must have five aspects:

1* A mental conception of a future
2* A conception of personal proximate causation
3* A conception of choice

4*A conception of will

5*An ability to prospectively compare the expected results of the proposed action to a set of 'best' secular purposes or to the ideals of a transcendental world view.

To begin, we need a definition of 'justice' to help frame the discussion. To wit:

"The subset of cognitively created and ethically based rules that prescribe the reparative consequences of specific, harmful social interactions."

Justice should also respond to harm in such a way so that the level of prescribed and applied consequences is proportional to the force applied in an action. Furthermore, not only the level of harm incurred from an action, but intent must also be considered.

When the personal emotions are allowed to alter the prescribed consequences, the result are specific subtypes of justice, such as revenge and mercy. The underlying common instinctual reactions to harmful actions do influence the logic of impersonal justice. But there are usually practical reasons given for the laws that govern the emotionally satisfying consequences. In the very large divergent groups, and subgroups, that comprise today's total world population, we find many different forms of applied justice. Not only that. If one considers the variables of three descriptions of action (harmful, neuter, helpful) times two types of C-L inputs (intentional and unintentional) times three types of I-E evaluations (positive, neuter and negative) one has at least eighteen different potentials 'causes' for any one action. This does not include other variables such as the concordance or dissonance between the two evaluative brain-mind systems or other situational factors. Furthermore, helpfulness and harm do not exist as a clear dichotomy. Enormous wisdom will be needed to create and implement a worldwide system, based on a rational, secular consistent ethic. First, there must be created the conditions for a common understanding of the human condition. This involves the continued examination of the historical and current systems of justice, comparing the prescribed consequences (law) in light of the common goal. This is a large order, that currently contains more hope than adequate implementation, as noted in the Epilogue. But current efforts should lead in that direction.

Justice, within one's subgroup(s), is one of the stated aims for almost every thoughtful person. Even a child will speak of fairness, a less structured concept than the implementation of justice. The latter is the active correction of perceived states of affairs which were caused by a lapse in following ethical rules (codified or not), as well as happenstance. Applied justice, in general, is therefore a response to past human behavior and its consequences. This term should not be used when responding to unwelcome outcomes from the activities of the inanimate and subhuman living world. Nevertheless, the response should be guided by beneficence. Empathy, compassion or self-interest may also initiate justice. The cases of non-human harm should lead to community effort with shared responsibility. "It's not my fault" is an inadequate response. It speaks of a lack of compassion, without which the will to act is not forthcoming. To fruitfully address these problems, the important concepts of free will, evil, and the applicability of a particular ethic to a specific situation must be explored.

15.2. Here we discuss the major underlying concepts of ethics.

Definitions of the functions of the brain-mind, such as cognition, instinct and emotion have been discussed previously in Part I to III, especially under "Ontology of mind" (See § 9.4 and Glossary) and are built upon below. As defined there, we have the Preconscious Adaptive Mechanisms (PAMs): simple reflexes, instincts, and habits. This is in contrast to the emotional system and conscious cognitive systems. The former is initially preconscious, but rises to the conscious level especially when the hormonal aspect is strong. The latter includes awareness of the models created by the imagination and problem solving by a use of logic. These abilities are additions to the autonomic NS which controls or influences the bodily functions such as heart rate, blood pressure, unconscious breathing, and so forth. All the subsystems of the NSs, and its totality, depend on multiple internal feedback loops and have variable impact on philosophical topics. Ethical rules are based on the evaluation of needs, satisfiers and values; terms, which received extended discussion in Part III. Although many of the pertinent concepts in the Wikipedia quote are derived from psychology, psychiatry, sociology, history, and economics, they can be usefully considered in terms of their impact on the subsets of concepts that are clearly of philosophical interest.

The above definition with its concepts defines the content and limits of the study of ethics and morals based on actual everyday understanding of the terms. They do not deal with what the rules actually are, or how they are to be applied. In his book "Ethics — Inventing Right and Wrong", J.L. Mackie 1977) makes this distinction. "Morality is a species of evaluation …. that is supposed to bring (something) about". (pg107). 'Something' is a personal ethical system that is meant, "… primarily to counteract the limitation of men's sympathies."

It is obvious that before one can expect ethical rules to lead to the valued outcomes of proposed normative action, one must make every effort to assure that the rules apply to reality, are logically used, and rely on information that is as objective as possible.

15.3 The Difficult Path to a Universal Humane Ethics and Justice

Justice is a frequent and strongly espoused ethical aim. However, if one is pressed to define its substance, this causes some difficulty. Partly this is due to the fact that this word has been used indiscriminately for several different subtypes of justice which are related but are quite different in action.

Retributive justice, such as "an eye for an eye", I will call "retro-justice", meaning both backward and outdated. It is often the impetus for further unethical action. It results from an imbalance between the C-L and I-E evaluations, with a preference given to the latter. A second, socially codified type of justice is legal justice, jurisprudence, which may be more thoughtfully applied. These two types of justice will only be discussed for contrast with the third. This is distributive justice, also called social justice, a topic whose best-known recent discussion was in the book "A Theory of Justice" by the philosopher John Rawls (1971). These three forms of justice seem to have little in common in terms of the kinds of problems they deal with, or the types of resolutions attempted. But if one looks at the negative, that is, when one asks what is injustice, the relationships of the types become clearer. In any case, any form of justice must fit in with a more general ethical system.

Injustice is perceived when a state of affairs or relationship is found to clash with an imagined state that is more equitable, the fairness of the young. All three types of justice are meant to undo, rectify or balance a real or perceived, loss or deprivation following some action. This leads to a discussion of what makes actions right and wrong. In this context, a right action results in greater equality and a wrong one in increased inequality of the human condition. The hard question is whether these concepts of right and wrong are arbitrary, or are emergent high level human concepts based ultimately on natural laws. Theological and mystical systems of correct action may include a planner-god or forces beyond those allowed by the P-laws. In these cases, the definitions of what is right and wrong is defined by diktat, unfazed by logical discourse. Following secular philosophers, including Lord Bertram Russell (1968) and Mario Bunge, I have found this scenario unproductive and unresolvable.

The question becomes, can ethics and therefore justice be based on understanding of the natural world, without revelation, using the models created by the imagination, as tempered by scientifically gained knowledge. I believe that given the advances in our understanding of the workings of the universe, including the human brain-mind, a more positive response can be given than even one hundred years ago.

Along with logical evaluations, the concerned emotional feelings of imbalance which demand social justice, underlie the meaning of the following poem which I wrote many years ago. I was a community pediatrician at the time and it will be unsurprising that it deals with justice for children, and also for the adults they become.

>Viewpoints
>
>The stars are shining, all unseen
>Behind the city's smoky screen
>City's child sees by the light
>Of neon sign the urban blight.
>
>Waves crash on distant shore
>Unheard by ears of child so poor
>Who hears, instead of gulls and terns,
>Sirens scream of death and burns.
>
>Is the world the same to these
>Who in the dark and dampness wheeze?
>To those who may enjoy, partake
>Of wondrous nature, pristine lake?
>
>Judge not harshly the child of need
>Who did not choose, where its seed,
>Carelessly, was quickly planted
>In a world so darkly slanted.

For me, this poem captures the yearning for the needed distributive justice its proponents have. It also serves as a warning against falling into negative judgements against those living in less fortunate circumstances, and from there into a retro-justice mode of thinking as well as an inflexible jurisprudence. Most written codes of jurisprudence tend to be inflexible systems because they are unwilling to make the attempt to take into a fuller account the complete circumstances. And, they allow for very little application of wisdom by a supposed impartial judging system. A secular, distributive system of social justice is more than an eye for an eye, a tooth for tooth. Neither is it about post hoc judicial punishment that does little to help those harmed.

In pondering the philosophical aspects of what such social justice encompasses and can be, three considerations, all equally important, yet equally fallible are:

1* Science, both the methods and the data driven models thereby gained
2* The current thoughts on this topic in philosophy
3* One's own imagination

Each of these sources contributes to the following results in different ways. The third is used whenever one thinks about the future or how things might be. The content of such imagination is limited only by one's gift (talent) in that department and one's experience, which includes the second consideration. Science adds the basis and delimiters of what is supportable as possibilities.

The concept of injustice is based on a perception of a state of affairs that needs to be addressed because harm was, is or might be caused. The three major forms of justice have different approaches in seeking redress and solution.

Retro-justice concerns itself mainly with revenge and it is basically egocentric and in essence a form of evil. Jurisprudence is, at worst, a primitive application of socially accepted revenge by the application of a set of rules, rules which may or may not apply to the inequity at hand. Nor are the attempts to make the victims whole a major consideration. Civil justice (civil court in the USA), while victim centered, is not held to a strict observance of fact and science. It often yields results and methods that are an overreaction to the lack of concern for the victims by the criminal justice system.

Social justice, however, concerns itself with the bigger picture. It asks itself not only about the underlying actions and the results, but concerns itself with how these conditions came to be. More specifically it is asking the questions; "What were the conditions?", "Is this an ongoing injustice?", "Are there more than one victim and perpetrator in this situation?", and finally, "Was the harm intended?". In a criminal judicial sense these all refer to physical or ownership harm. But what of emotional harm, character assassination, false witness and so forth. These emotional harms and harms to one's standing in a community are often ignored or left to be dealt with on a personal basis. At best (or sometimes worst), the civil courts are involved. To understand the full situation, one cannot only respond to what occurred, but what were the effects.

To do this there must also be some recourse to sympathy, which can be defined as empathy and its evaluation by cognitive means. The capacity to feel empathy is not universally shared, and has no legal standing. The ability to apply cognitive evaluation is similarly not equally found or applied. If a judging person does not have typical amounts of these capacities, both the victim and perpetrator are not likely to receive full justice. For the capacity to feel empathy and its derivative, sympathy, science supports both genetic as well as environmental factors. The ability to personally pursue full social justice is like a talent, which like all talents must be supported, honed, and practiced.

This emotional capability is illustrated by the biblical story concerning the "Wisdom of Solomon". Solomon was confronted by two women, both of which claimed to be the mother of the same child. Solomon used his capacity for empathy to help him understand what must be done. (One must remember he did not have genetic testing at his disposal.) By suggesting that the child be cut in half and evenly distributed, he used empathy to ascertain that the true mother would never allow that, despite the loss of her personal relationship with the child if it was given to the wrong woman. By the reaction of both women, he was able to ascertain the truth, and thereby administer true justice to mother, child and perhaps the imposter. We are not told what happened to her. Unfortunately, juries and judges are not chosen according to their wisdom. It is not clear whether this is even possible. This is more likely to happen in a small community where all the players know each other for many years. Because of this, in a modern large community, judgments should not be final or irreversible.

Empathy and wisdom do not seem to play an obvious role, if one considers the example of contract law. However, if one does not understand the whole picture, by only examining the written words without knowing the conditions that led to that particular contract, injustice, may be done. If the contract is not fair on its face, one must investigate whether fraud, misrepresentation, pressures or even ignorance was involved. The phrase "buyer beware" does not speak of justice, but rather the rule that "might makes right". The source of might in this case includes purposeful deception by withholding pertinent information. Innate abilities, upbringing, life experience and knowledge of social norms need to be addressed for justice to be forthcoming in these cases.

Finally, harm perpetrated by societies on out groups is a special problem. This is unfortunately a very common problem in even the smaller societies. As observed by Albert Schweitzer, tribal groups tend to have very little empathy outside their own group. He describes episodes where there was a lack of willingness, even refusal, to help carrying stretchers on which were patients from another tribe. It took him years of tireless example to change the local ethic at Lamberéné. This all-too-common lack of empathy for the other emphasizes the need for a humane ethic that must be formulated, discussed and supported on a broad, worldwide basis.

There are several underlying components of distributive social justice. Among them are:

1• The physical and psychological needs of an individual are not chosen, but inherent and due to variable factors.

2• Within and across groups, values, needs and their satisfiers can be approximately, if incompletely, normalized. But only if all parties feel and understand the respective needs and resources of the other.

3• Justice must balance the needs and satisfiers between any individual or group. Disparities must be dealt with ethically and compared to a single, high-level goal, such as subsumed under Schweitzer's phrase "Reverence for Life".

To achieve this stage of humane ethical development will be a difficult and long process but never complete in all the details. But if one desires full justice the attempt must be made to increase wisdom in the service of the goal of promoting a world where humanity is in balance within itself and with its environment.

15.4 Subtypes of Applied Ethics: Medical, Legal, Business, and Other Groups

There are many subgroups of ethical guidelines that refer to, and are based on, specific types of power. A list of "xxx ethics" would replace the "xxx" with: medical, business, science, political, military and other professions. They are individually concerned with the specific powers and their potential harms associated with the group referenced. An example was given for medical ethics in § 13.4, but here we will only deal with the commonalities. In the subsequent sections, some specific, high profile ethical problems, such as abortion, euthanasia, capital punishment and war, are then discussed in light of the commonalities.

Specific professions have formal lists of ethics, promulgated by committees of their respective organizations. College courses, seminars and many books and articles have discussed and analyzed the many respective subtypes. There we often find the urge to create absolute rules when creating ethical guidelines, fearing the supposed laxity and 'slippery slopes' of mere rules-of-thumb. Unfortunately, this often leads to unwarranted condemnation of specific actions in specific circumstances. Also, inappropriate (unfair) responses are then instituted. The issue of specificity is however neglected, despite its importance, when ethical rules are applied. How to balance the simplicity of absolutes versus general rules-of-thumb, when considering the specifics of a case, is where the professional ethicist or philosopher can demonstrate and apply wisdom.

The slogan "think globally, act locally" sums up these divisions concerning the general and the specific. This phrase is currently mostly associated with environmental concerns, but can be equally applied to a much broader range of issues. But, as noted previously, these concerns and the responses to them, are more and more important to philosophical discourse. "Think globally" addresses the idea that our means and goal should be global, that is universally applicable. "Act locally" follows from the fact that our means are usually limited to a much smaller sphere of influence. The sphere of responsibility is of course directly related to the power of each actor. The term, locally, therefore encompasses home, community, country, and regions depending on individual or group resources. A will to act is also needed. The power available is in turn dependent on the financial, material, formal (organizational and governmental), persuasive, and charismatic resources. Before addressing each source of power, a further general aspect of the term is noted. Power, no matter how great or small, can be metaphorically compared to gravity. The effects of both are inverse to the distance, physical and temporal, between actor and the affected. It is also additive, yielding a vector outcome, whose direction and impact depend on the number and relationships between the forces.

It is easy to see that the greater the availability and use (or threat of use) of any power(s), the larger the impact. This differential in the level of harm or helpfulness will necessarily vary greatly depending on the distance between actor and that which is changed. The first three sources of power: financial, material and formal are commonly perceived and appreciated. By formal I mean to include the power of status associated with political, military and organizational hierarchies. The relative power of bishops versus priests, Senator over local council member, and CEO over employee is usually great — and obvious.

More subtle, but usually ordinally quantifiable, are the powers associated with the persuasive and charismatic talents. That charisma lends itself to persuasion is clear. But in certain situations, the perception of someone being a "knowledge leader" is also useful in promulgating harm or helpfulness. The combined vector of the effects will be composed of the application of all the available powers. Furthermore, any power will be amplified by the combined power of followers, acolytes, and co-conspirators, however small they may be, individually and relatively. This is especially notable in the realm of political and religious persuasion where sometimes the group overwhelms the power of the original leader.

Finally, and not unimportantly, there are the psychological, biological and physical characteristic of the affected entities. These influence the receptivity or resistance to the applied power. The combination of applied powers and the inherent properties of the affected lead to the actual outcome of the interaction. It is another reason why rules-of-thumb should be the start of any ethical discussion rather than an absolute dictum.

This conclusion leads to the following, less discussed problem of ethics: are the rules-of-thumb applicable in dealing with those whose power is used for harmful, or worse evil, purpose?

15.5 Levels of Power, Its Effects and the Subsequent Application of Justice

In every ethical interaction there are levels of power and effects – both harmful and helpful. These levels do not affect the application of an ethic, but rather the amount of effort needed to provide justice. This is why stealing is equally wrong whether it is the taking of a penny or a million dollars. But the actions needed to provide justice change accordingly. The former may only need a reprimand in order to educate the perpetrator of the value of the ethic. Larger thefts may need stronger education, reparation of the victim of the thefts, and other measures for incorrigible thieves. What those measures are will depend on each situation. The person who steals food because society does not share; the psychiatrically challenged thief with kleptomania; and the incorrigible, anethical personality will all need different remediation. The situations are highly variable, and a proper just response becomes more and more difficult the larger the group. Mostly this is due to the much greater variety of influences in a given situation. The lack of personal knowledge concerning the actors also interferes with the appropriate evaluations. All these difficulties do not justify the mostly punitive reactions of both personally and judicially applied justice. Advice in this area is where a rational philosophy could make a large impact, always with the caveat of the underlying hypognostic state of our best models of action. The ethical rules-of-thumb should rarely be the final reaction, but only a starting point.

15.6 Life and Death

The phrase, *"Death, the high cost of living"*, from Neil Gaiman's graphic series (1993), is a telling summation of the relationship between the two concepts. The formal cause for the process of life is treated in the chapters concerning its ontology. Death, the irreversible end of that process, is its necessary conclusion. Various theological systems try to soften the psychological blow by positing an afterlife or a cycle of recurrence. The scientific realist, who acknowledges his/her hypognostic epistemological concepts as based on observable data and logic, cannot confirm these hopes. In contrast, believers in mystical systems justify their belief on the basis of unconfirmable concepts. This replaces the uncertainty of the unknowable with confabulations that prop up their certainty concerning otherworldly entities. Based on shareable information concerning what reality may contain, only realism can claim defendable justification for its concepts and related beliefs. The actual properties of our universe are not determined by belief, rather they are the basis of our incompletely confirmable understanding (model system) of said universe. This TRUE (see § 6.1) universe has nothing to do with the "truths" of agreed to definitions or systems of logic and maths. The latter only provide workable methods and icons that make conciliant understanding possible. Nor are our 'true' models incontestable if better data is found.

Life is a process; death is the cessation of the same process. Most precisely the unqualified term, death, should be reserved for the cessation of the living process of a cell. The cell being the basis of all other groups of cells that we call alive. These are clonal groups. All individual organisms that ever lived on this earth are based on clonal groups of cells, often supplemented by symbiotic relationships with other cells or groups of cells. In the cases of their individual demise, we should talk of clonal death. Logically, the word death should be modified to correspond to the modified uses of the word life it refers to.

There are many uses for the term 'life', each of which is a species of the general generic term. One speaks of a love life, the intellectual life, the artistic life, the life of the community, college life, and subhuman life among others. The respect for the latter was the central theme of Albert Schweitzer's writings and will not be further examined here. It is these other non-cellular "lives", but all cell dependent, that give extra meaning to humanity. It is these qualified, "dependent lives" that give the subjective value to individuals and that lie at the heart of the discussion of the value of the individual "human life."

Human beings, like all other multicellular organisms are a structured collection of cloned cells. Such a multicellular collection is produced asexually from one cell to which it is genetically identical (except for incidental mutations that occur during the creation of the structure.) It is the subject of embryology and other developmental sciences. These clonal collections interact with other living beings in symbiotic, cooperative and parasitic relationships, that are not central to the discussion of an organism's life and death. The death of clonal organisms is but the sum of all the individual cells dying. The individual life of all the component cells rarely ceases in an instant, but over time. Such clonal death must therefore be defined differently from cellular death, when not instantaneous, like in an atomic blast. The pertinent question is: "When is the process of clonal death irreversible and when are the cellular relationships disrupted beyond repair? Whereas cellular life is the basis all the other forms, and clonal lives are legion, these two biological processes are not what drives ethical debates. It is the actual and potential loss of the dependent lives that seem to matter. All living cells and clonal collections of cells have, depending on level, an ascending collection of tropisms, reflexes, instincts, and habits. These mechanisms are powerful defenders of the living process. However, we cling to dependent emotional life and intellectual life, emanating from the combined actions of the cells necessary to support a typical human brain mind. These two are but the most forceful of all the specifically human dependent lives, only possible from the emergent properties of the typical functioning human brain mind. With this understanding a discussion of 'intentional human caused death of individual H. sapiens beings' follows. For the remainder of these discussions the term "intentional death" will stand in for the longer phrase.

15.7 Purposeful death: Capital Punishment, War, Abortion. Euthanasia, Lethal Neglect

The willful endings of individual life come in several varieties which can be further subdivided. The first division is between the ending of a human clonal life versus other beings large and small. This section will deal with the former. The ethics of willful death of the latter is, of course, dependent on the ethics of the former, but is in most cases less emotionally fraught. Another division is the death of individuals versus groups. The former includes capital punishment, abortion, euthanasia. These can further be differentiated based on developmental age and inferred psychological potential. The latter includes war, and the mass neglect or actual deprivation of basic necessities of defined cultures and communities. Wars and lethal actions against groups of others are more perverse in that they make not individuals, but rather sacrificial groups the recipients of random death. These differences call for a nuanced discussion even though the outcome for any individual is the same: the end of the clonal life.

These mechanisms of these intentionally caused death differ in the underlying intentions of the assaulters and the characteristics of the victims. The intentions and characteristics are the main parameters that define the ethical aspects of the actions. A related, but separate issue is the amount of suffering perpetrated before final unconsciousness and clonal death are reached. The suffering caused is an additional harm that should have its own unique ethical evaluation and consequences for the assaulter. A short discussion of the subtypes follows.

The imposition of capital punishment, a form of revenge and murder, questions the basis of any ethical system that pretends to have consistent absolute answers to the rightness or wrongness of any of these actions which lead to clonal death. It is claimed by some that it has a deterrent effect on others, but if so, the deterrent effect is very weak and poorly focused. For example, the promulgation of religions has not been stopped by the murder of martyrs. Neither has the martyrdom of revolutionary leaders stopped further revolutions. The execution of the instigators of aggressive wars and mass murder of citizens has not stopped the repeated violent efforts of dictators and madmen. The kinds of ethical responses (justice) to evil are dealt with in the next section. Without a clear outline of appropriate justice, ethics are hollow and arbitrary.

The case of aggressive war involves a combination of direct, personal violence and mass impersonal massacre – and needs more ethical discussion than has been forthcoming. Most modern violent "tribal" conflicts, of which war is but a larger form, impose the certain danger of clonal death on those that would reap very few (if any) of the benefits of victory compared to the initiators of the conflict and their clique. As such, its instigation is inherently an ethical conflict of interest. Defensive, warlike efforts can only be justified if the goal is to stop the violent conflict. Once the defenders have the upper hand, all efforts must be made to stop the killing. The capture of the instigators must then follow. But, as in capital punishment, the emotional urge for revenge by the victors causes only more unnecessary death and should be seen as murder. Unfortunately, once mobilized, the violent emotions are not easily abated. The negative social outcomes are also due to the expanded potential of the human brain-mind to exact revenge many years after an initial confrontation and harm has occurred. It is a major underlying cause of capital punishment and prolonged violent conflict. That does not, however, ethically justify such precognitive passive responses by the victorious leaders, whose task is to lead both the defensive war and find the path toward peaceful coexistence.

But it is also true that humans are, objectively, different in their potential. The combination of the biologic and, especially, mental mechanisms toward self-preservation has led to the concomitant lethal neglect of "the other" and the grievous state of today's man-made environmental destruction of this earth. Mass extinctions, global climate change stresses (most yet to come) are a result. The ability to change the local environment to meet biological need has, with the resulting population explosion, completely unbalanced the normal natural processes that allowed for many-fold niches for tens of thousands of species. This will be of great detriment to future generations of all forms of life, including H. Sapiens. Also, the unnecessary mass starvation and general human misery is another outcome of human overpopulation, aggravated by a lack of ethical consideration of distant others. This outcome is, today, potentially reversible with preconceptual birth control. How low the human population would have to be, to preserve the positive advances and better control the harmful outcomes, is unknown. Halving the current population of earth over several centuries would be a good start, but the methods would be fraught at best.

For both abortion and euthanasia, a central issue is how, and whether, to differentiate between fully functional individuals and those who are functionally incomplete. Although there are shared commonalities the details differ considerably. For both actions an often-ignored consideration is the ability and willingness of the pertinent community to support a "life worth living". When such help is inadequate, misdirected or, withheld then the biologic and dependent life of a clonal self is suboptimal or even decidedly harmful. The ethical goal is not just cellular life but the ability of each individual to partake as much of the dependent lives as necessary to reach the minimal perceived worth of the clonal life. The final question is: "At what point is the level of such dependent lives of sufficient value to prolong the clonal life for the affected?" The greater emphasis on the importance of dependent lives is seen in how we treat the same issues in plants and animals. It is the lesser potential (or absence) for these "life worth living" mental abilities that leads to clearer evaluations of the balance between a prolonged life with pain, suffering and lack of purpose versus bringing about an intended death with the least amount of pain.

With this in mind, we will now proceed to these individual issues. In the background it must be remembered that all living forms of life will end at some point with the death of the full set of specific clonal cells. Lives are not saved in perpetuity, but only prolonged. Does duration always trump quality?

Abortion can be natural, induced purposefully or as side effect of another action. It is, biologically the interruption of the process that leads from the fusion of a gamete to an organism that can function outside the mammalian womb. It is a special case of embryocide (as of this time a neologism which has also been suggested to the Collins dictionary in 1918). The term also refers to the same interruption of progress of a fertilized egg or the interruptions of marsupial development toward the independent organismic state. Holistically it can refer to any organism that reproduces by sextual means.

In most cases, ethical discussions of abortion are limited to the human domain. But if one takes such a limited view, one must define the ethical difference between human abortion and other willful embryocide. This hinges on Schweitzer's "Reverence of Life", the process, and the care we feel toward any particular instantiation of life. In his discussions he gives preference to the balance needed in the earthly ecosystem, without negating the care for the individual living things. It recognizes the inevitability of death, as an actual necessity for that balance. In most philosophic systems, individual humans take precedence because the instinctive preservation of the self is very strong, or from religious conviction. The understanding of the needed balance is then neglected.

The fetus has, as yet, no personality that has been formed by an interaction of genetics and the non-uterine environment. That is what makes normal birth one clear dividing line. This unfinished organism has the potential for causing future harm and good. The inferred potential tends to lean toward the latter unless prenatal investigations strongly suggest the former. Otherwise, there can be no way to decide whether a fetus will or most likely be, in a positive or negative way, typical or atypical. The atypical negative can, unfortunately, cause much more harm than the atypical positive can have beneficent impacts. Can an Edward Jenner (smallpox prevention), Jonas Salk (polio vaccine) or Martin Luther King Jr (justice for the minorities and poor) make up for the depredations and mass murder instituted by the Stalins, Hitlers, Napoleons and Genghis Khans – to name but a few well-known exemplars? There is no way to measure the number of lives prolonged by the former, now and into the future, or the undercounted premature deaths ordered or facilitated by the latter. Building a healthy world and society is always much more difficult than their destruction. But what of the, much more common, impact of the typical potential persons? Therein lies the conundrum – the individual or the whole community? Modern western society has favored the former, eastern societies the group, and leaders preponderantly their self-interest. But the additive effects of the mass of all the typical human beings is now more harmful than the most destructive political, nationalistic, xenophobic and self-interested warlords of the past. It should be noted that the purposeful or accidental start of a nuclear war could change that calculus. But that is a "what if" that cannot be foretold.

Neither abortion or euthanasia can be justified as a source of population control. Firstly, such a goal is not the underlying psychological reason its proponents and defenders have. It is but a minor added benefit. Can these two actions then be justified, and under what circumstances?

Abortion, by eliminating the potential future life of a developing fetus, causes harm, the loss of that self-limited potential. Wherein lies the overbalancing benefit then? It is of most direct benefit to others; parent, family and community – in specific cases. Specific cases! No generalizations can meet the test of blanket justification. Most often the reason to abort a fetus is because it is expected to take away resources for already developed beings. This would be a much less persuasive reason if the larger community distributed the cost of a new being beyond the mother and close family. But that is rarely the case. If every "pro-life" person adopted as many children as necessary to provide the psychological and material support every young child will need, and society as a whole taxed itself to provide the material support for a healthy productive life, then almost all the negatives of unwantedness and poverty would be minimized. But that is not the case. A natural "experiment" was the "orphanages" of the Nicolae Ceausescu regime in Romania up to 1989. (For a description and references concerning the ensuing horrors see the Wikipedia article "Romanian orphans".) It is arguable that for many of these tens of thousands of children it would have been better if they had never been conceived in such a society. It would have been better if these (counterfactual in retrospect) potential individual had not been left to the mercies of an uncaring world. Would an abortion have been a mercy?

Each and every pregnancy needs to be evaluated for all these impacts. This is however only incompletely possible; hypognosticism notwithstanding, there will be real impacts forthcoming. Lack of omnipotence does not excuse lack of thoughtful action. Today, the future of those fetuses, whose anatomy shows by ultrasound, or their genetics foretells a certainty of irreversible disability (and a dependency that is not offered the full help of society) is one of incomplete dependent life opportunities. In most societies, including the rich western ones, they and their families are left to fend for themselves by the body politic. Charitable resources have not, and never will be enough to lighten the load in most cases.

Since society is not willing to take on the responsibilities, it should not have the last say on this issue. It is, for most mothers-to-be, a very difficult and lonely decision – and they should not be second guessed by those who show insufficient ethical interest in the outcome. In converse. abortion should also never be imposed on a mother who knows the potential love and anguish that will be forthcoming for the length of her life, and the prolonged life that the fetus is the beginning of. Biological lives are never saved for eternity. They are prolonged or shortened, made easier or more difficult, more enjoyable or more stressful. Ethical actions will try to find a balance, and that means a case-by-case evaluation of all the factors. There are no absolutes that fit all occasions.

Euthanasia is similar in that it deals with the end of the life of one organism and some of the considerations of abortion apply. For the former, the terms handicapped or disabled have been used to justify the end of that life, as it is. The first term refers to an increased difficulty to fully function independently. The second refers to the (near)total inability to function independently, in specified functional domains including the potentials for the dependent lives. All the domains have a range of abilities from absent to adequate to exceptional. Again, each case is different. Whereas the potentials of an unborn infant are unclear, the potentials of "a life worth living" in an afflicted child or adult is clearer because more information is available. It is the potential for the dependent lives that come to the fore. If the possibility of actively giving or being aware of love is gone, the wish or ability to be a part of society is near zero, then the case of prolonging the clonal life is very weak. Again, if society will not exert every effort to maximize the dependent lives, it should have little to say compared to those who can still inform us of their wishes or those who act as their conservators.

An application of absolute ethics in these cases is a mirage. Ethical guidelines are the best we can formulate. That, and deep thought concerning each individual case. The answers to the questions of prolonging a clonal life will be made with less doubt and more compassion.

Suicide, self-inflicted clonal death, is often a failure of a community. It speaks either of lack of awareness of a person's inner torment, or much worse, inflicted torment that makes any dependent lives not worth the prolongation of the clonal self. It is self-induced euthanasia, but usually without thought to ethics. Assisted suicide is clearly a form of euthanasia, somewhere between the difficulties of euthanasia and suicide. It comes about because of the physical inability of a person to end his/her own life, and the actual inability to provide adequate relief by the community.

In all cases determinations of the proper ethical way forward can only be discovered by a close examination of the situation. As with all C-L processes this supposes enough time to gather all the pertinent facts and only then initiate an action to end a life, or provide more support.

15.8 Is an Ethical Response to Evil Power Wise?

The logical answer to the question is "yes", though the emotional inclination will often be "no".

Schweitzer (1936, pg. 234). paraphrasing Ghandi. says that ".... the use of force does not become ethically permissible because it has an ethical aim, but in addition it must be applied in a completely ethical disposition."

What then should be the ethical social response to all harmful actions? In all the lesser cases, the actor may benefit from education, retraining of habits or removal from the conditions that are likely to give rise to the possibility of further similar harmful action. This question is the ultimate test of how we characterize evil, ethics, and wisdom. To review:

Ethical action is the result of weighing the imagined balance of harm and helpfulness resulting from specific intentional action or passive intentional inaction. Ethics can never guarantee this balance because of the hypognostic status of the imaginative process.

Evil is the intentional active infliction or intentional passive acquiescence to harm, for harms sake, or for personal material or emotional satisfaction.

Wisdom is the outcome of personal experience, education and empathy. Together these attributes narrow the uncertain gap between a guaranteed, ideal outcome and an achievable one by using the resources of a brain-mind that has striven to maximize these three properties.

Taken together, it follows that evil needs to be resisted, but not at the cost of more harm and less helpfulness than the evil action entails. After all, ethics demands action that is calculated to be, on balance, helpful. But to resist and if possible, overcome evil power one must usually apply equal or greater countervailing power. The potential sources of this counteractive force include the same ones as those that underly the harmful actions and intentions. By ignoring whatever wisdom one can muster, and letting the emotions of apathy, fear or revenge rule over our actions will lead, overall, to even greater harm, and/or a failure to lessen or eliminate the unopposed source of the harmful outcome. It is a reaction to the symptoms not the causes.

One other unwise set of actions is to simply attack the means of the evildoer, rather than the source. The most common example of this is in war, where the burden of harm falls mainly on the pawns of leaders and their cliques.

The most effective, but ethical, action would be the aim of eliminating the power of those leaders and their cliques with the minimal foreseeable unwanted side effects. The number of variables involved would require a group of the very wise. Unfortunately, the adequate power to counter the initial evil is too often held by other leaders and their respective cliques. These initiators of harm, too often, avoid wisdom in the name of efficiency, revenge or lack of personal risk.

Is this prospect, of mainly fighting against the obvious symptoms without addressing the causes, inevitable in the future of our species? Pessimistically, and unfortunately, in the foreseeable future, the very likely answer is yes. The Epilogue proposes some sources of a hope that a better outcome is feasible and achievable over time. It will require the combined efforts of many ethically driven, if hypognostically limited, brain-minds. That is, wisdom must be aimed for in a much larger proportion of humanity.

Epilogue: A Rational Hope

Hope is an emotional reaction that offers a counter to despair when the likely potential future and its meaning promise harm to our desired purposes and ethical goal.

A hoped-for, significant, natural, and unforced Darwinian change in the human genetic profile, leading toward an increased average inborn talent and desire for a humane and worldwide ethic, are dim. Eugenics is not the answer because at that scale there will be inevitable misuse. That leaves an increase in the production, dissemination and accumulation of realistic but emotionally consilient ideas that foster a desire for wisdom. In other words, a scientific approach that includes strong psychological data. This should then lead to a program of widely disseminated education based on the best understanding of human nature, at its best and worst. (Sapolsky, Robert M. 2018)

This will require a threshold of minimal understanding of the centrality of reality-based models before individuals and societies demand such an education for themselves and their children. Until then, unsubstantiated, but seductive models based on wishful thinking and wignorance, will be promulgated for the benefit of powerful, though selfish and parochial, persons and their cliques. Philosophers must come to see the wisdom of becoming active in the process of achieving this understanding, along with scientists, reformers and activists. The aim is to include scientific realism into the popular culture. I personally would also welcome theologians, whose purposes include and prioritize improvements for the living rather than for unsubstantiated afterlife benefits.

** GLOSSARY **
With Pertinent Cross References to SECTIONS

THE MEANING OF WORDS AS USED IN THIS BOOK

"What do you call alive, exactly?" Chang said. "It's kind of a matter of semantics."
    Quoted by Jeremy Deaton in NBC MACH article 10/14/19

"The search for definitions is never easy, particularly in such fields as the social sciences [and] Gnosticism...Nevertheless, the present chaotic conditions warrant an attempt." www.gnosis.org/whatisgnostic (S.A. Hoeller)

"When I use a word,' Humpty Dumpty said in rather a scornful tone, 'it means just what I choose it to mean — neither more nor less.'
'The question is,' said Alice, 'whether you can make words mean so many different things.'
'The question is,' said Humpty Dumpty, 'which is to be master — that's all."
    Lewis Carroll, *Through the Looking Glass* (1871) in "Humpty Dumpty "https://en.wikipedia.org/wiki/Humpty_Dumpty). 2021

The concept comes first, the symbol for the concept, -usually a word, comes after. (See "Meaning" sec. 4.6)
                    * * * * * *

▲   Important related word in the Glossary.

§   Notable section(s) of the book where the word is used.

←   The following is a description, definition or explanation.

→   See Chapter, section number or indicated entry for more detail

In the description a CAPITALIZED word is a related term in the Glossary.

A

ABDUCTION ← Coined by C.S. Pierce. A form of ▲ INDUCTION that has been called "Inference to the Best Explanation." (SEP, 2021). It is consistent with the limits inherent in the concept of HYPOGNOSTICISM.

ABSOLUTE ← Without exception: for ethics → § 14.8

ABSTRACT (noun) ← An incomplete part of the whole it refers to with a definable relationship to that whole. (Verb) The process of noting only a part of the whole.

Physical and mental → §§. 5.6,5.7

ABSTRACTION (verb) ← The process of producing an abstract. This process may be mental, artistic, conceptual, a form of communication, etc. Each separately named process will have different mechanisms and the consequences of the process should be clarified in any serious discussion that uses the specific term such as "mental abstract", "artistic abstraction" and so forth.

→ §§ 5.7,5.8

ACTION ← When changes in energy causes changes in the relationships of specific units of MATTER/ENERGY.

AESTHETICS ← The field that deal with the principles of beauty, its correlates and antonyms. It concerns itself with fully subjective evaluations with only a modicum of COGNITIVE input. → §§ 12.2 -12.4, 12.10

AFFERENT NEURON ← Those nerves that react to stimuli from non-neuronal sources. Compare with e*fferent neurons* that send information to non-neuronal cells and *central neurons.* (see §§ 3.11 and 5.6 for details).

AGATHONISM ← A proposed highest GOAL. From γαθός, "self-preservation with a social conscience". → § 10.5 (Mario Bunge, 2003).

AGNOSTICISM ← Transcendent: In mysticism and theology reserved for the necessary unknowability concerning transcendent entities.

← Secular: The acknowledged uncertainty of all mental models of reality. In conjunction with transcendent agnosticism, it forms the WORLD VIEW of HYPOGNOSTICISM.

← Agnostic atheism: "is a philosophical position that encompasses both atheism and agnosticism. Agnostic atheists are atheistic because they do not hold a BELIEF in the existence of any deity, and are agnostic because they claim that the existence of a deity is either unknowable in principle or currently unknown in fact." (Wikipedia). ▲ ATHEISM

Glossary

AIM ← An imagined future state of affairs. Often used interchangeably with PURPOSE and ▲GOAL. Near synonym is DESIRE which has a stronger emotional component. ←In ethics more than one goal leads to the certainty of contradictions. → § 13.6

ALIVE ← Having the property of LIFE.

ALTRUISTIC ← A behavior when one organism gives a resource to another without obvious recompence. It is a form of neutral relationship initiated by the "giving" organism. In a sentient being it may be partially SYMBIOTIC because of the positive self-esteem generated. Near antonym of PARACITIC.

ATHEISM ← a MODEL of the universe that axiomatically excludes a transcendent godlike causative force. Neither it, nor Theism, can be verified and therefore cannot be discussed on a scientific basis.

ASPIRATION ← A GOAL or AIM that includes the emotional aspects of DESIRE.

ATYPICAL ← Arbitrarily the top and bottom 10% of a distribution of values. The percentages may differ if the distribution is significantly different from a normal curve. ▲ TYPICAL.

AXIOM← Premise, Assumption. Used especially in science, mathematics and systems of logic. In mathematics it is usually *a priori*. In science it may also be the result of INDUCTIVE reasoning that finds common causes and relationships in all cases investigated. If it leads to contradictions or non-consilience when used in practice, or as part of logical relationships, the axiom, the logic or the data must be questioned.

AXIOLOGY← The study of value including the concepts of good and evil. →*Chapter X*. For the impact on ETHICS see introduction of Part IV.

* B *

BEAUTY← An AESTHETIC evaluation. → *Chapter XII*

BEING ←" In philosophy, **being** means the material or immaterial existence of a thing. Anything that exists is being. Ontology is the branch of philosophy that studies being. Being is a concept encompassing objective and subjective features of reality and existence "(SEP, 2021).   Can be divided into:
      MLT = Modeling Living Thing
      NLT = Non-Living Thing
      RLT = Reactive Living Thing
  → *§ 7.7 including Table I*

BELIEF ←MENTAL MODELs based on: scientific data (atoms), personal experiences (apples), unverifiable imagination (apparitions), and received form others (any of the first three). → *Chapter V*

← FAITH and CERTAINTY. All three underlie KNOWLEDGE, an undifferentiated term for mental models. → § 5.13

BENEVOLENCE ←Intentional support of a GOAL or AIM. ▲ HARMFULLNESS & HELPFULLNESS, also ▲ GOOD & EVIL.

→ § 13.3. If the goal itself is EVIL then its support is not benevolent by definition.

BRAIN ← A collection of nerve cells and the supportive structures that is the material underpinning of a **BRAIN-MIND**. Subunits, from smallest to largest, are neurons, ganglia, larger structures such as amygdala and substantia nigra that are in turn part of more diverse anatomic structures named midbrain, occipital lobe etc. Local and diffuse connections between all these levels form functional systems. Together their activity is the BRAIN-MIND.

BRAIN-MIND ← Energy dependent activity of the physical, animal BRAIN. Contrast with the simpler activity of an artificial system (computer).

→ §§ 2.2, 4.1, 5.11, 9.13). Humans have a brain-mind, and it is currently the most complex one known.

* C *

CARDINAL ← Cardinal numbers indicate a specified numerical quantity. ORDINAL Numbers indicated relative order or position in a series, such as less than, greater than, equal and so forth.
Used to differentiate relative values. → § 9.11

CATALYST ← A catalyst is a substance that changes its structure during its involvement in a process, but reverts to its initial state at the end. The concept is most commonly used in chemistry, but can be applied to other similar interactions as well.

CAUSE-EFFECT ← A description of MATTER/ENERGY relationships. Causes are temporally prior to the effects that result. Most effects on complex matter, that is change in level or type of its energy or spatial relationships have multiple causes, some major and many contributing minor ones. → CHAOS, MECHANISM.

← Concerning Free Will. → § 13.11

CERTAINTY← An absolute BELIEF. Because of our HYPOGNOSIC limits, it is usually unwarranted when speaking of mental MODELs. Exceptions are tautologies and systems based on arbitrary AXIOMS of mathematics and logic systems.

CHANCE ← A common term for ▲ PROBABILITY, usually used in an ORDINAL way.
It is invoked to describe the likelihood of CHANGE, CREATIVITY, DESTRUCTION. It is the alteration of the properties of a defined item of MATTER/ENERGY.

CHANGE ← (Noun) A difference of properties of a stated set of MATTER/ENERGY at different times.

CHAOS ← Ontological: In the universe as a whole, it implies that there can be no axioms that must apply to everything within it, making the CAUSE-EFFECT concept of questionable importance in explaining constant relationships of MATTER/ENERGY.

← Epistemological: CHAOS THEORY- the idea, and fact, that small differences locally, can lead to wide differences in outcome down the cause-effect action tree. This makes the calculations of prediction in complex systems impossible over long distances of SPACE/TIME and large iterations However; this incalculability does not negate the concept of cause-effect.

← Social: the lack of or non-enforcement of community rules or laws.

CLONAL ORGANISM ← ▲CLONE. Five types with identical DNA are:
1. Single independent cells (bacteria, amoeba, most of algae).
2. Undifferentiated group of adherent cells (some parts of life cycle of algae, slime molds).
3. Differentiated adherent cells from a single zygote (most plants, fungi, animals).
4. Formed by fission of adherent cells as in identical twins, triplets, budding (Hydra, yeast), fragmentation (starfish, annelid worms).
5. Parthenogenic from unfertilized ova (some insects, fish, amphibians and reptiles).

All multi-organ humans are either type 3 or 4. However, many multicellular organisms are also highly dependent on SYMBIOTIC relationships with other species.

CLONE ← A multicellular organism or single cell, or group of organisms or cells, produced asexually from one ancestor or stock, to which they are genetically identical. A multicellular organism is a collection of adherent cells all arising from a single ZYGOTE but displaying different properties that together make the collection, animal or plant, viable. Single cell organisms, bacteria, amoeba etc. are independent of each other. Does not necessarily include the community, but rather the individual, of a species where highly similar sisters and parthenogenic may form subgroups that specialize (ants, bees). All life is composed of single living cells, but DEATH occurs at individual, community and species level and each type should be differentiated. → § 15.6

COGNITION ←The mental activities that perceive (react to) stimuli, both external and internal, which then is followed by consciously evaluating those stimuli independent of the separate emotional evaluation. Together they may lead to conation, that is the sense of purpose and desire that powers the will to action. Cognition may, however, also use the emotional evaluation as an input. It does not include the Preconscious Adaptive Mechanisms. ▲PAMs

COGNITIVE DISSONANCE ← When the two major mental MODELS (I-E and C-L) are noncompatible. This may be due to erroneous data and/or logic, conflicting memory and the different mechanisms that lead to those models. This dissonance can lead to defensive reactions such as ▲WIGNORANCE, and avoidance of thinking about its causes. Constructively it can lead to more consilient ideas, the search for better information by many means.

COMMUNICATION ← Intended transfer of information. Intention presupposes a brain that can foresee into at least the near future and anticipate the reception of the information. Stimuli that are part of tropisms, instincts and chemical environmental interactions that are part of a life's forms of reactive and feedback systems are not communication. Also, non-living MATTER/ENERGY can only have interactions. → §§ 8.1 -8.4

CONCEPT ←▲ MENTAL MODEL. It is the current model based on past experience. It can be modified by PERCEPTS over time in the processes of IMAGINATION. Initially (in a newborn and after) concepts are formed by repeated stimuli from the environment and one's senses. It is then available to be retrieved into the subconscious workings leading to its conscious appearance in short term memory.

← Conceptism – a neologism equivalent to another neologism, "IDEA-ISM". Not the same as the archaic "conceptism" of Spanish mystics.

CONFABULATION ← the mental process, based in part on FANTASY, that fills gaps of information in order to create an internally consistent mental mode that reaches consciousness. Historically its use was limited to the process caused by pathologic memory loss, but it also includes the process of how concepts are created by belief in data, unsupported by experience, in non-impaired populations.

CONSCIOUSNESS (noun), CONSCIOUS (verb) ← A subset of the cognitive activity of the mind, whose activity is reflected in SHORT-TERM MEMORY. → §§ 5.15, 7.5. An animal that has this property is said to be conscious ▲ SUBCONSCIOUS which is composed of the UNCONSCIOUS and PRECONSCIOUS.

CONSCIOUS–LOGICAL (C-L) ← Those circuits of the brain that are responsible for creating realistic MODELs of the world, especially the PREFRONTAL CORTEX. Compare with the INSTINCTUAL-EMOTIVE (I-E) system. → *§§ 9.11 and Table II*

CONSEQUENCE ← The source of MEANING. The outcome of any ACTION or interaction.

CONSILIENCE(n), CONSILIENT (adj). ←
"In science and history, consilience (convergence of evidence or concordance of evidence) is the principle that evidence from independent, unrelated sources can "converge" on strong conclusions." Wikipedia (2021). → *§ 1.3*

CONTENTMENT← A positive mood that does not reinforce any urge for change. HAPPINESS. → *§ 11.7.*

CREATIVITY ← A source of CHANGE. The output of IMAGINATION. It is not applicable when the change is due to repetition (habit) or unmodified recall from memory.

CULTURE← The sum of commonly held beliefs, values and standards as well as their expression in practice (arts, science and religions), of a defined group of people. → *§ 6.9*

D

DATA ← Singular, datum. Unprocessed measurements, enumerations or other OBJECTIVE description. Also, the stimuli impinging on a life form. Once organized by some system of logic, specific sites on a cell or by a group of cells (especially the nervous system) it becomes INFORMATION.

DEATH ← The cessation of life. Unqualified it should only refer to the cessation of the ▲LIFE processes of a cell. CLONAL DEATH indicates the death of all cells of a CLONE. Clonal Death occurs with the total death of any single cell species (extinction) or individual multicellular life forms formed from one original zygote (typical ORGANISM death). This includes the death of all plant ramets →ZYGOTE or identical twins. The latter are especially interesting from a philosophical perspective because individually they are a form of organism death, and together, clonal death.

DEDUCTION ← A form of reasoning leading to a specific result when axioms and data are combined using logic. If the axioms, data, or logic are faulty the conclusion is likely to be faulty. Related to INDUCTION, and ABDUCTION.

DEFINITION ←Phrase or equation, composed of SYMBOLs, that limits the MEANING of a concept or its ICON. It is recursive because other symbols, icons or definitions are needed to fulfill this task.

    ← The Meaning of Meaning. → § 4.6

    ← In Ethics. → §§ 13.2, 13.6, 13.7

DEISM ← Belief in a creator God who set down the laws of existence, but is not intrinsically good and is transcendent rather than imminent. This entity neither hears or answers prayers. Its existence itself is speculative. ▲ SKEPTICISM

DEPENDENT LIVES← One speaks of a love LIFE, the intellectual life, the artistic life, the life of the community, college life, and subhuman life among others. These are the lives that makes biologic life worth living for a human whose biologic life is not threatened. → § 15.6

DETERMINISM ← The concept, or philosophical stance, that all changes over time are governed by a CAUSE-EFFECT mechanism. → Free will §§ 13.11-2

DESIRE ← An AIM that has a large input form the emotional brain.

DESTRUCTION ← A change leading to the loss of organization. Leads to an increase in ENTROPY. Pertains to the full range of organization from, at least, the level of atomic structure to the organization of matter needed for life.

DIMENSION ← The three dimensions of space and that of time. The number of parameters to describe a point in a material object. The higher number of dimensions found in the equations concerning quantum physics have no use on larger scales. Also used to refer to parameters of a concept such as ethics.

DNA ← Deoxyribonucleic acid. The macromolecule that contains the information that programs the constant activity of a cell and also heredity. It is, metaphorically, a blueprint of most organisms. Simpler forms use RNA (ribonucleic acid), a slightly different structure. → § 3.7

DREAM← PRECONSCIOUS IMAGINATION. It may be potentially stored as a new memory if it occurs just previous to consciousness that comes with awakening. It is an important part of long-term memory consolidation.

DUALISM ← ▲MONISM.

E

ECOLOGY ← The study of the interaction of living systems with each other and with the general physical environment.

EGOTOPIA ← The personal ideal. Details will change with time and it may be an instantiation of a more general, group UTOPIA. → §§ 7.4, 11.2

EMERGENCE ← the axiomatic process that creates new possibilities by changes in MATTER/ENERGY relationships. → § 1.1. It is central to the ontology of the philosophical school of ▲ EMERGENTIST MATERIALISM.
(https://www.nbi.dk/~emmeche/coPubl/97e.EKS/emerg.html)

    ← Emergence of life and mind → §§ 2.1-2

    ← Emergence of the evaluative capacity → Chapter IX, § 10.8

EMERGENTIST MATERIALISM ← "The version of ▲ EMERGENTISM that holds that, although all real existents are ▲MATERIAL, they are grouped into different ▲ LEVELS" (Bunge, 2003). (Original typesetting) It is a subset of SCIENTIFIC REALISM.

EMOTION ← Neurologically preconscious, centered in the Limbic system, ▲BRAIN, but involving hormonal systems as well. It interacts and influences the conscious process and usually is included in one's awareness. In turn an emotion may be triggered by a conscious thought. It is differentiated from MOOD, which is a weaker form that is often not in conscious awareness. Because of the hormonal component it is not included in the Preconscious Adaptive Mechanisms. ▲PAMs.

EMPIRICISM ← Philosophical model in which all mental models are ultimately derived from sense experience. How this intersects with mathematics is unclear. Rationalism allows for inherent mental processes to contribute to the formation of these models. But extreme rationalism denies the importance of sensory input. A scientific realism allows both the senses and the inherent capacities of the regional and neuron-based brain anatomy to make important contributions to the formation of mental models.

END ← As in "the end justifies the means" is used synonymously with ▲GOAL.

ENDOCRINE ← A system of internal stimuli using a subset of chemicals called hormones. Some of these signals are important in responding to and modifying NS activity, especially the emotional circuits.

ENERGY← The property that allows the performance of work (change) on another object. The conservation of energy is one of the axioms of science. Gravity is the potential energy of mass. Energy alone is a property of electromagnetism. ▲MATTER/ENERGY.

ENTITY ← A collection of relationships that functions as a distinct whole.

ENTROPY ← Loss of usable energy.

ENVIRONMENT← Of an organism, whether internal or external to a living organism, can be a useful differentiation. This follows the same distinction made for STIMULUS. The _internal_ environment consists of local, cell to cell, paracrine interactions via molecule release. Action at a distance uses the nervous system and circulatory system. The later uses chemical messengers, such as glucose or more complex molecules such as the hormones, of the ENDOCRINE system. Both are needed to maintain a stable equilibrium, called homeostasis, of the multicellular organism. Similar, but of lesser importance to NS activity, are the juxtacrine interactions which are local cell to cell signals that require contact.

The _external_ environment in this context includes all entities beyond the physical limits of the organism that may or do influence it. This includes organisms that populate the skin and the gut.

One can also speak of the environment of any other entity.

EPISTEMOLOGY← Study of ▲KNOWLEDGE. It concerns how and what one can learn. (The subject of Part II).

ETHICS and MORALS ← Rules of human behavior. "**Ethic**: refers to the set of moral values and principles which, taken collectively, guide and influence the life of a group or, less commonly, of an individual. **Ethics**: (in the plural) is the code of behavior considered appropriate to a particular group, especially a group defined by its occupation or profession." (https://english.stackexchange.com/).

A major component of Philosophy comprising Part IV of this book.

← Definition of: → § 13.2

←Humane ethics → §§ 13.10, 14.6, 15.3. An ethic that takes BENEVOLANCE seriously.

←Absolute Ethics. ← and IMPOSSIBILITY → § 14.8

ETHOS/PATHOS/LOGOS← Forms of persuasion. **Ethos is** an appeal to ethics, and it is a means of convincing someone of the character or credibility of the persuader. **Pathos** is an appeal to emotion, and is a way of convincing an audience of an argument by creating an emotional response. **Logos** is an appeal to logic, and is a way of persuading an audience by reason. Persuasion is important in ethics. → § 15.4

EUKARYOCYTE← A cell that contains DNA inside a nucleus. A result of the evolution of life forms. → §§ 3.7, 3.9

EVALUATION ← "Evaluation is a systematic determination of a subject's merit, worth and significance, using criteria governed by a set of standards." (Wikipedia (2016). There are two main types, CONSCIOUS and EMOTIONAL. → *Part III, esp. Chapters IX and XII as well as §§ 10.10, 10.11 and 11.11*. The mental activity that determines ▲ VALUE.

EVIL← Often used in the phrase "Good and Evil" ▲ GOOD & EVIL. → § 13.3

EXISTENCE ← A concept important to philosophy. Having the potential to affect change in another existent. Therefore, the minimal need for a universe, to "experience" existence, are two QUANTA.

← Material E.: Anything that has the properties of MATTER/ENERGY.

←Conceptual E.: A mental property that is composed of patterns of matter/energy. Such patterns are relationships in space and time

← Relational E.: A property between two quanta (or any of the constructs that they are in). Space and time both depend on this. Relationships and SPACE/TIME are mutually dependent on each other by definition as in $E=MC^2$, where C is the distance (in space) is divided by time.

EXPERIMENT ← A method used in science. It is the collection of data by observation in a controlled setting where parameters of interest are varied in a systematic way. It is the best way to test a HYPOTHESIS or THEORY, but is not always possible. This is especially true in astrophysics and the biological evolution that occurred in the past. Here careful observation and organized gathering of data is called for.

* F *

FACT ← Mental fact: A certain believed MODEL of reality. → BELIEF.

← Defined fact: a linguistic connection between a concept and the words used to describe it.

For example: Triangles have three sides. This is a fact because a triangle refers to a structure composed of the connection between three points on a plane connected by lines. This interconnectivity of words with concepts can be traced back for every word used in a language.

Proper names also fall in this category as do definitions in abstract systems such as mathematics.

Compare with ▲ TRUTH.

FAITH ← An absolute, often untestable, BELIEF; may be based on any experience. It is a necessary approximation of CERTAINTY that underlies intended actions meant to effect a specific consequence.

FANTASY ←A MENTAL creation that is only partially based on REALITY which may be supplemented by CONFABULATION. Its relationship to reality is therefore uncertain. → § 5.4. If it is known to be unsupported by referral to reality, it is called fiction. It is a product of the imaginative processes of a BRAIN-MIND, but it must be reviewed by the C-L system to assess its relationship to REALITY.

fMRI ← Functional Magnetic Resonance Imaging. A technique that allows near real time images by the use of computerized analysis of the reactions of molecules to a magnetic field. It is used in medicine, neuroscience and psychological studies, mostly by mapping blood flow differences in various tissues under differing situations and between subjects. A specialized form of MRI.

FREE ← Without restraint. Subsets in uses of the term consist of the source of restraint or its presence or absence from that source. The lack of definition of possible sources lead to the fraughtness of the use of the term FREE WILL.

→ Levels of Freedom → § 13.12

FREE WILL ← The concepts behind this term, often pointing to freedom from external cause, are very many and highly disputed. It is important in ETHICS. → §§ 13.11-.12 ▲ WILL. It is on the spectrum of: Fatalism — Determinism (of Idealism) — Determinism (of neuro-realism)— Indeterminism — Full Free Will (Freedom of the Will).

* G *

GENE ← In technical use a distinct sequence of nucleotides forming part of a chromosome, a string composed of nucleic acids and other molecules. A gene may be a blueprint where the sequence determines which a protein is synthesized. Some also carry the information for other cellular functions. Important in the field of Genetics and Darwinian inheritance. It is the main site of mutations.

GOAL ← An imagined ultimate END result. Only one ULTIMATE GOAL can be held if an ethics is to be consilient. AIM, DESIRE, PURPOSE, and MEANS (noun) are often seen as "goals", but should be placed in a subservient category to the goal. All are important in ethics → § *13.6*. Non-contradicting aims and their associated actions must also be investigated. A goal shares the non-instrumental property of intrinsic value. → § 10.1. AGATHONISM is such a goal.

GOD(S) ← Hypothetical TRANSCENDENT beings, usually with extra powers such as omnipotence, magical abilities, immortality etc. not available to a member of *H. sapiens*.

GOOD & EVIL ← Terms loosely used to indicate opposites. This is inexact. "Good" is a property that supports an aim or goal – and does not need to be intentional. "Evil" is always intentional. Good (HELPFULNESS) and bad (HARMFULNESS) are a better pair as are benevolence and evil. The pair is often used (inappropriately or vaguely) in ETHICS. → §§ *13.3,13.12*

* H *

HAPPINESS ← a stronger feeling (emotion) than contentment and usually shorter lived as well as less sustainable. An EMOTION. → §§ *11.6-7*

HARMONY / DISHARMONY ← In music two or more notes that have a pleasant/ unpleasant sound. Mental harmony is a metaphor indicating lack of / or presence of fear, doubt or difference in the values derived by the C-L and I-E systems. → § *11.8*

HEDONISM ← "**Ethical** or evaluative hedonism claims that only pleasure has worth or value and only pain or displeasure has disvalue or the opposite of worth." ("Hedonism" in SEP).

HELPFULNESS & HARMFULNESS ← An evaluation, either retrospective or of the imagined effect of an action. When an action is intentionally harmful it may be evil, if intentionally helpful it may be benevolence. → § *13.3*. ▲ GOOD & EVIL.

HOPE ← secular prayer. Wishing for a future containing desired aims and goal by the intercession of probabilities that are against the odds. It is the mental I-E system's response to probable negative outcomes as imagined by the C-L system.

HUMAN ← A member of the Homo genus. Modern humans are often given the name of *Homo sapiens sapiens*.

HUMANE ETHICS → An ethic that that considers the combined typical needs and abilities of cognitive beings. → §§ *14.6, 15.3*

HUMANE PRAGMATISM ← An ethic whose actions are derived from combining the foreseen consequences with a benevolent will. → § *13.10*

HYPOGNOSTICISM ← A neologism. → § *1.1*. It combines the theological agnosticism, coined by T.H. Huxley, with MATERIAL REALISM. It is that vast area between omniscience and absolute ignorance. It acknowledges the incompleteness of data leading to nature-based axioms and MODELs that contribute to the ONTOLOGY of Material Realism. It goes beyond SKEPTICISM as a positive statement of possibilities. It also includes ▲AGNOSTICISM, which by its nature is based on a posited, but unconfirmable, transcendent realm.

HYPOTHESIS ← A MENTAL MODEL (MM) that seeks to explain a set of data using induction. A THEORY is a firmer MM based on extensive testing of the model (by further detailed observation or experiment).

* I *

ICON ← a SYMBOL, often pictorial.

IDEA-ISM ← A suggested name for an epistemology that states that a MENTAL MODEL can only be approached in reality, never reached. It is consistent with the concept of HYPOGNOSTICISM. Alternative could be "conceptism".

IDEALISM vs. REALISM (Philosophical) ← Classical: Plato vs Aristotle. The former places the mind as the source of reality, the latter that the mind mirrors or abstracts reality as MENTAL MODELs. Classically, Platonic ideals include the perfect circle → § *5.12* or personal EGOTOPIAS → § *7.1* and social UTOPIAS. Realism proposes that the mind, by imagining the future as based on current concepts, can then change the future possible reality by initiating actions that affect the
PRESENT. (See IDEA-ISM above).

Glossary

IDENTITY ← the uniqueness of an entity. Subtopics include:
  ← Ship fable of Theseus – physical self – structural "blue print"
  → § 10.4
    ← Personal Identity. → § 10.4
Dual personality, split brain, and psychological self are psychological variants that are pertinent and present difficulties for a simple definition of identity of self.
    ← Clonal self. DNA pattern identity → § 10.5 → CLONE
    ← Twins; conjoined → § 10.6

IGNORANCE ← The lack of DATA, or its derived KNOWLEDGE. It is often unavoidable because of the limited amount of all data a mind can hold. HYPOGNOSTICISM is therefore the appropriate epistemological stance. ▲ WIGNORANCE.

INSTINCTUAL-EMOTIVE (I-E) ←Those circuits of the brain that are responsible for the automatically derived MODELs of the world, especially in areas of the limbic system (hypothalamus and amygdala). See also CONSCIOUS-LOGICAL (C-L) system. → § 7.7 and Table I

IMAGINATION ←   The process by which conscious mental MODELs (concepts) are ultimately formed. The mechanistic details are as yet not clear, but it involves input from the interior and exterior SENSES, MEMORY, UNCONSCIOUS and PRECONSCIOUS activities of the expanded neocortex. → §§ 5.9, 11.1

IMPERATIVE ← Authoritative command. Kantian Categorical and hypothetical imperatives → §. 14.8. Also, of vital importance → § 4.6

IMPOSSIBILITY ← In a logical system and in science a result that contradicts an axiom. More loosely, any result that is so highly improbable that it would only occur once (if at all) in a time longer than the age of the universe. In Ethics → § 14.8 the later definition would hold. Incorrectly used for the concept of unlikely or low probability in a conceivable amount of time.

INDUCTION ← a principle, generalization or AXIOM derived from DATA and LOGIC versus ▲ DEDUCTION.

INFORMATION ← DATA that has been organized by a system, mathematical or mental, making it potentially useful to a living organism, specifically an MLT.

INSTINCT ← One of the Preconscious Adaptive Mechanisms. ▲ PAMs. It is a complex reflex mechanism. → § 10.7

INTELLIGENCE ← A property of BRAIN-MIND that allows MENTAL MODEL formation.
    ← The unwarranted expansion of the term→ § 13.7
    ← Misuse in term "artificial intelligence" → § 2.2
INTROSPECTION ← Process of MENTAL MODEL creation using only input from MEMORY without concurrent external stimuli.
ITERATION ← Repeated process over time.

J – K

JUSTICE ← A judgement of what balance of relative HELPFULNESS AND HARMFULNESS is appropriate in a community. There is retributive, legal (jurisprudence), and distributive justice. It is the application of an ethical system. This complex subject is discussed in Chapter XV.
JUSTIFY ← The act of rationalizing the acceptability of a MEANS because it leads to a desired GOAL. The rationalization may be logical and based on the facts of necessity or be unsupportable, because of ignorance, or worse, WIGNORANCE.
KNOWLEDGE ← The core subject of Epistemology (Part II of book). A MENTAL MODEL of reality as a conscious thought. Those aspects of MEMORY that are not currently in our conscious awareness are part of the UNCONSCIOUS and PRECONSCIOUS brain-mind. → § 7.5.; Contrast with ▲ BELIEF.
KURAL, THE ← Also Tirukkural, a classic Tamil "text consisting of 1330 short couplets of seven words each .... divided into three books" concerning virtue, wealth and love. "Considered one of the greatest works ever written on ethics and morality, it is known for its universality and secular nature". ("Kural" in Wikipedia).

L

LANGUAGE ← A set of symbols and rules used to communicate concepts and their intended meanings. If the symbols do not point to the same concepts, then the meanings will be different for those that use the symbols and those that receive them. That is why the clear definition of supposedly common symbols (words, graphics, icons, sounds, hand signals and so on) is critical to communication and discussion. → Prologue, § 4.6.
    Also, its intersection with what it means to have KNOWLEDGE → § 5.2 and its importance to WISDOM. → § 8.2-4

LAWS OF PHYSICS ← AXIOMS derived from scientifically garnered data using logic. This book includes EMERGENCE as a necessary component to explain forthcoming, new, consistent properties when entities interact. The term →P-LAWS is used for that expanded definition.

LEVELS OF FREEDOM ← indicate the actual number of forces that contribute to a particular outcome. The lower the level of freedom, the more contributing factors there are to a result. (This is not the same as degrees of freedom in statistics).

LIFE ← "Life – bounded, macromolecular, hierarchically organized and characterized by replication, metabolic turnover, self-repair and exquisite regulation of energy flow; constitutes a spreading center of order in a less organized universe." (Based on Grobstein, C. (1965)) → *§ 3.3*

   ← "Life on the sun" → *§ 2.4*
   ← Emergent properties of life → *§ 3.4*
   ← Value of life → *§ 10.3*
   ← Ethical purposes of life → *§ 11.5*
   ← A life worth living → *§ 11.9*
   ← Life and Death, DEPENDENT LIVES → *§ 15.6*

LOGIC ← A system of rules about both CARDINAL and ordinal relationships. The former deals with numbers. Mathematics is a subset of logic. UNCONSCIOUS mental EVALUATIONS are based on a poorly understood system of inborn rules.

   ← Predicate logic. A learned system of rules that is used to find the TRUTH of statements. A precise statement is called a PREDICATE in the academic discussions.

* M *

MACRO ← Prefix denoting larger than the usual size of the concept in question. Compare to MICRO.

MAN ← A phenotypical male human adult. Linguistically paired with woman, a phenotypical female human adult. Often used generically for HUMAN because no word exists in English for the German MENSCH other than PERSON. The latter is used in this book. Does not address gender, the self-identified social role.

MASS ← The property of matter, usually measured in grams, that offers resistance to change in speed or location in space.

 MATERIAL ← Ontologically anything that has the properties of MATTER/ENERGY as modeled by the scientific method.

MATHEMATICS ← A subset of ▲ LOGIC. → § 2.11

MATTER/ENERGY ← the duo that comprises the totality of the physical universe. Each quantum can combine with others, and their previously formed combinations, to form new and different classes each defined by their content and complexity. Einstein's formula $E=MC^2$ shows the basic unity that underlies the convertibility of the two forms. Matter has the property of MASS and occupies space. Energy is the property that drives change or does work. The speed of light, "C", involves SPACE-TIME which are relational properties.

MEANING ← In philosophy and linguistics refers to a relationship. But this is incomplete. Meaning is more properly understood as all the actual effects of a CAUSE-EFFECT action. → § 4.6

MEANS ← The proposed or actual CAUSE/EFFECT energy transfer to attain AIMS and the GOAL. → §§ 13.8, 14.7

MECHANISM ← In the CAUSE/EFFECT relationship of scientific realism there are three inputs. They are the laws of motion, the P-laws of matter/energy and emergence.

MEMORY ← Changes in the PHYSICAL structure of the brain due to new internal input. The new patterns are then accessible to consciousness and used by the IMAGINATION process. It is the encoding and storage of information that is retrievable by various CONSCIOUS and UNCONSCIOUS BRAIN-MIND systems. The MECHANISMs involved are only partially understood.

 ← Long-term memory ← (LTM) Changes that are permanent for hours or a life time.

 ← Short-term memory ← (STM) Changes used to process current information and is needed for consciousness. It is composed of changes that last seconds to hours. It receives data and information from the LTM and immediate sensations.

 ← Working memory ←Used in assessing the environment and for solving individual pieces of a puzzle from the data in STM.

MENSCH ← The German word for human being or person. It is gender neural and can used as a way to indicate a gender-neutral term. In Yiddish it acquired the meaning of an exemplary person. The word MAN will be consciously avoided if the male gender is not meant to be explicit. "S/HE" is used in this book for the general pronoun.

MENTAL ← Any property of a BRAIN-MIND.

MENTAL MODEL ← composed of relationships of MATTER/ENERGY in a brain. It is a phenomenon resulting from the interaction of both internal and external (of the body) stimuli with the anatomic and physiologic aspects of a complex nervous system. It is a special subset of physical **MODEL**s. The resulting concept are formed using the creative processes of the IMAGINATION.

METAPHYSICS ← Has multiple definitions. The simplest is that it is the study of being and cosmology. The latter has become the prevue of astrophysics, while the former is a specific ontological stance of, most commonly, realism and idealism. Idealistic metaphysics is often at odds with a scientific realism and cosmology because it allows for non-OBJECTIVE study by includes the FANTASIES created by the imagination of individuals.

MICRO ← Prefix usually denoting size below the unaided vision. Other uses are analogs. Compare to MACRO.

MICROBIOME ← Microscopic living environmental influences such as bacteria in the gut.

MIND ← The activity of a brain ▲ BRAIN-MIND or other complex decision-making structure (? plexus-mind). A possible parallel would be the metabolism/cell or body "duality". <u>The mind-body problem</u> is a linguistic one – mixing an activity with a substantive entity in the term. For short discussion of best alternative, ▲**EMERGENTIST MATERIALISM** (See *Bunge, 2003)*. May be subdivided into the SUBCONSCIOUS (unconscious and preconscious) and CONSCIOUS mind.

MLTs ← MODELing Living Things versus NLTs and RLTs. ▲BEING → § 7.7

MODAL ← From the statistic use of the word mode. The numerical value of a characteristic that has the most members in that group. May by used instead of "average", "common" or "normal" when referring to such a characteristic because the other three terms are also used pejoratively. ▲TYPICAL, the term I mostly use, is used to describe the larger central group of values. Philosophy of the "common MAN" would be therefore referred to as the philosophy of the Modal or TYPICAL human being. It indicates that we are not speaking of unusual, rare, or outlier cases.

MODEL ← An abstraction representing an instantiation. Besides a MENTAL MODEL there are other common forms: Architectural models and computer models are examples of secondary, external instantiation of a mental model. Fashion "models" are the carriers of a physical example of the mental model the fashion designer had in mind.

MODELING (mental) ← Activities of the brain that gives rise to CONCEPTs that are available for evaluation by either the emotional or cognitive systems.

MONISM ← The principle that all of reality can be described by a single basic and irreducible building block. → §§ 1.2, 4.6. It is the basis of atomism. In current physics there are two entities derived from mathematical manipulation of data concerning Matter/Energy properties. These entities are subatomic quarks and electrons with properties described by String Theory. This therefore forms a materialistic form of dualism. This is contrasted with idealistic *Dualism and Pluralism*. The former allows for some form of materialism and a transcendent or immanent, non-materialistic substance. It is the basis of Idealistic metaphysics. The latter allows for multiple, incompletely defined basic units of reality. The processes of ▲EMERGENCE make both of the latter views unnecessary, but allows for all the variation seen at the macro level of existence.

MOOD ← Related to EMOTION. It is generally weaker in strength, longer in duration, and much less focused on an object. It is not a Preconscious Adaptive Mechanisms. (PAMs).

MORALS ←Personal view of right and wrong. → ETHICS.

MRI ← A static imaging technique using very strong magnets. ▲ fMRI.

MYSTICISM ← "In modern times, "mysticism" has acquired a limited definition, with broad applications, as meaning the aim at the "union with the Absolute, the Infinite, or God". (Wikipedia). It is also applied to the belief in unmeasurable energies. → § 7.3

N

NATURE ← Synonym for REALITY. In the phrase nature vs. nurture, it means genetic endowment vs purposeful interactions meant to influence the development of the young of the species. The either/or of the "vs" is especially inappropriate for the mental activities of a human being where an "and" is really the case. → *§ 9.1*

NECESSITY← An absolute requirement for a result to be forthcoming.

NEED ← It is the perception of NECESSITY and often used as a less stringent synonym. It can be the result of either/or the C-L or I-E mental systems. It is a determinant of ▲VALUE.

NERVE ← Synonym for NEURON.

NERVOUS SYSTEM (NS) ← a complex comprised of autonomic, instinctual, emotional, cognitive, and intellectual subsystems. NEURONs are the functional cells with the rest being necessary as support for proper function of the BRAIN-MIND.

NEURON ← A specialized cell of a NERVOUS SYSTEM that gathers inputs and may then transmit an electrochemical impulse as a result. → § 3.11

NLTs ←Non-Living Things. Versus MLTs and RLTs. ▲ BEING → § 7.7

NORM ← Synonym of STANDARD, especially in ethics or behavior.

NORMAL ← ▲ TYPICAL. → § 4.4

NOUMENA ← plural of NOUMENON; used by Kant in his philosophy of knowledge → § 4.8; and PHENOMENON.

O – P

OBJECTIVE ← Data: Observation compared to a physical standard not colored by feelings or interpretation. It is limited by the type of sensor, mechanical or biological. SUBJECTIVE observation adds emotion, interpretation.

ONTOLOGY ←▲ Part I "The basic building blocks of reality and their interaction". Includes the P-LAWS that control EMERGENT properties. The latter allows us to speak of the ontology of mind. → §§ 4.1, 9.5. Not same as cosmology which is study of origin and constitution of Universe.

ORDINAL← ▲CARDINAL

ORGANELLE ← A structure in a cell that is separated from its immediate surroundings. This accomplished by a membrane or because molecules form as phase-separated liquid droplet. (See https://www.the-scientist.com/features/these-organelles-have-no-membranes-65090). Both allow increased rates of chemical reactions. For relationship to DNA → § 3.7. **ORGANISM** ← A functional unit of life consisting of one or more cells.

PAMs ← **P**reconscious **A**daptive **M**echanisms. This includes cellular reaction, reflexes, temperament, personality traits, talents, instincts and habits. It excludes emotions and moods because they involve the hormonal systems and have a greater integration with cognitive nervous systems. → § 10.9. ▲PRECONSCIOUS.

PARADOX ←A result from seemingly acceptable premises and logic, that is senseless, logically unacceptable, or self-contradictory. Either a premise or the logic is in error in a system with universal laws. →§§ 10.6, 14.8

PARASITIC ← A relationship between two organisms when one living thing takes a resource from another without equal compensation. It does not necessarily cause harm to the "giving" organism. ALTRUISTIC and SYMBIOTIC are near antonyms.

In ethics → § 14.2; In life and death → § 15.6

PERCEPT ← A MENTAL MODEL that is the product of a current eternal stimulus and previously formed CONCEPT. Perception is the verb for the action.

PERFECTION ← Totally without fault.

PERSON ← Used as the generic for a HUMAN or MENSCH.

PERSUASIVE ← The ability to change the viewpoint of another using ▲ETHOS, logos and pathos. It is the civilized method versus violence or threat thereof.

PHENOMENON vs. NOUMENON ← Kantian separation of our mental models and perceptions (the phenomenon) of reality from the latter (the noumenon) which is the underlying reality which we can never fully describe. The mental models of SPACE/TIME and MATTER/ENERGY, are themselves noumena.

PHYSICAL ← Composed of MATTER/ENERGY. All forms are NOUMENA.

P-LAWS ← The axioms of science, also called the LAWS OF PHYSICS, plus the concept of lawful EMERGENCE.

PLURALISM ← See MONISM.

POLYSEMOUS ← A polysemous word is one that is used for different concepts in different contexts, and worse, in the same context. This Glossary tries to address that important issue.

POSITIVISM ← an philosophy that claims more than scientific realism promises because it denies the limits of HYPOGNOSTICISM, namely the inherent uncertainty of human MENTAL MODELS (KNOWLEDGE).

POSSIBLE ← CONSILIENT with the P-LAWS. An evaluation using the C-L rather than the I-E system of the BRAIN-MIND.

POTENTIAL ← POSSIBLE emergent outcome, but not a necessary future given time, the current reality of a system and all possible future interactions.

POWER ← The property of having energy and context to change the future.

← Use against evil → § *15.8*

PRAGMATISM ← In striving for a result, taking the possibility and probability of success into account. Expounded by John Dewey (especially in his educational theory) and C.S. Pierce when examining whether a statement is meaningful. → *§ 13.10*. It denies the propositions of moral Absolutism which ignores context. The latter is the basis of many theological ethical systems and that of Kant.

PRECONSCIOUS ← Those workings of the brain that are not currently CONSCIOUS, but are available to consciousness if activated (by yet unknown mechanisms).

A subset of ▲SUBCONSCIOUS → *§ 7.5*.

PREDICATE ← In logic "a *predicate* is the formalization of the mathematical concept of statement".   In language "The predicate of a sentence is a portion of it which makes a claim about the subject" (Wikipedia). In finding TRUTH by a syllogism → *§ 6.1*

PREFERENCE ← A conscious ordinal evaluation of a possible future situation which would be/will be chosen if possible.

PREFRONTAL CORTEX ← The area of the brain behind the forehead and above the eyes.

PRESENT ← A point in time between the past and the future. If time is composed of quanta, it is part of reality. If not, then it is only a concept of current instant in IDEALISM. Mentally it is a subjective small range of time encompassing seconds to years depending on the context.

PRINCIPLE ← A conclusion reached by INDUCTION or used as a basis for a DEDUCTION, but less strictly investigated than an ▲AXIOM.

PROBABILITY ← Technically a CARDINAL, mathematically derived number. Loosely may also be used in an ORDINAL context. → *§ 2.10*

PROKARYOTES ← cells without a nucleus. → EUKARYOCYTE.

PROPERTY ← Any POTENTIAL of a system. It may be part of a model system or undiscovered. → § *2.10*

PURPOSE ← the imagined, consciously chosen future. A mental MODEL that gives rise to AIMS, which are made functional by MEANS. A GOAL is the ultimate purpose. → *§ 8.6*

PYRRHONISM ← Early Greek skeptical philosophy. A counter to dogmatism.

Q – R

QUALIA ← SUBJECTIVE conscious experience. Necessary for self-awareness.

QUANTA ← (singular, quantum) A non-divisible entity. Usually associated with physics, as the reality underlying all MATTER/ENERGY. Whether or not time or space are made of quanta has not been determined. Important to any ONTOLOGY.

RANDOMNESS ← Does not indicates a lack of causality. Rather it is a lack of predictability, but like the toss of a coin its probability may be calculatable to a high degree. → § 2.10

← And Free Will→ § 13.11

RATIONALISM← ▲**EMPIRICISM**

REALISM ← Philosophical stance that there is a mind-independent existence of a whole universe composed of MATTER/ENERGY, but denies transcendent entities. MENTAL MODELS, which are real configurations of a BRAIN-MIND, are concepts for the real things, their properties and relationships, ▲REALITY, but also of many things that do not describe any possible instantiations. Numbers do not exist, but refer, ultimately, to existence and nothingness – themselves a relational pair. Again, the models are real states of individual minds, but do not refer to transcendent or other instantiations of zero and one.

Most fantasies are the creation of the imagination that combines entities and their properties in ways that cannot exist according to the P-LAWS. This includes REIFICATION which is not a possibility in this philosophical system. ▲IDEALISM.

REALITY ← Absolute versus personal. When used without a modifier then absolute reality, a synonym for baseline ▲TRUTH, is implied. The confusions arising from the use of unmodified general terms such as reality, truth, environment etc. can usually not be avoided unless the context is absolutely clear (that is, there is no possible misinterpretation by the TYPICAL brain-mind). The idea that reality can be different for different people is due to the confusion between the noumena of Kant (the real world) and the individual MENTAL MODELs which are derived by our interaction with it. The latter can be usefully compared without predicating that a different reality is the source of the models.

REDUCTIVE FALLACY ← Oversimplification of causal relationships. Also known by the Latin phrase *"reductio ad absurdum"* that is reduction to absurdity. In Ontologies that do not recognize the central importance of EMERGENCE it is used in an attempt to prove the impossibility of a given result. If the premises and logic of a situation, in concert with tracing the connection from them to the result are in concert with the actual result, it is a good sign that the system is close to correct.

REFLEX ← A simple motor response that is the result of an unconscious cause-effect loop consisting of the firing of an afferent neuron due to a stimulus, sent to the cerebellum or spinal plexuses, resulting in a firing of an efferent neuron to a muscle causing tension or movement.

→ A COMPLEX REFLEX has more steps and can involve higher centers. → § 4.9

REIFICATION ← When non-material or imaginary "spirits", ghosts and pure energy are treated as if they had the properties of matter. This cannot, axiomatically or by any data driven cause-effect mechanisms, be supported in REALISM.

RELATIONSHIP ← It is an emergent property that compares or describes the consequences of the interaction and/or their relative placement in SPACE/TIME of two or more non-identical components of the universe. It is a component of MEANING.

RELIGION ← A highly developed mental construct the places us in a definite relationship to eternal time and infinite space, or beyond, and often with an imagined transcendental plane. Less organized are various forms of mysticisms and confabulations that posit unfounded "things" outside the MATTER/ENERGY totality of this universe and posit a relationship of humanity with them. There is no data that supports those transcendent ideas. However, they are often the source of ethical systems that are valuable for defining human relationships with REALITY.

RLTs ← Reactive Living Things. Versus MLTs and NLTs. ▲ BEING.
→ § 7.7

S

SATISFIER ← One of the useable MATTER/ENERGY inputs that increase the probability (gives it the VALUE) of lessening a NEED. → §§ 10.1, 14.1. If the VALUE becomes negative the input would be an aggravator.

SCIENCE ← A method. Can start with an observation followed by an orderly search for data. This set of data may then lead to a first hypothesis using INDUCTION. The addition of further data may then cause a change or improvement of the previous HYPOTHESIS and possibly to a THEORY.

Accidental observations may lead to a pathway of: Questioning of meaning of observation, then organized data acquisition, followed by development of a hypothesis leading to further data acquisition and, hopefully, at some point allowing the construction of a well-formed theory. The process of this never-ending loop is never complete. The use of controlled experiment, where possible, is the other major method used to gather further data. Examples:

A. Fleming (discovery of antibiotic properties of Penicillin);

C. Darwin (development of the theory of evolution);

M. Curie (discovery of radiation);

And many Astronomers (planetary science, black holes, quasars etc.). In subjects such as history and archeology gathering data in an orderly and unbiased fashion is akin to observational science. This gives credence to the idea that they too are a form of science.

SECULAR ← non-religious or non-MYSTICAL. SCIENTIFIC REALISM belongs in this set of world views. It excludes transcendent MENTAL MODELS because there can be no experimental verification, and controlled observation have never found any consistent findings that stand up to thorough investigation.

SCIENTIFIC REALISM ← The philosophic ontologic stance stating that data derived by scientific methods are the only reliable basis for our MENTAL MODELs (Bunge 2003). → § 1.1. It is the concept that the whole universe can in principle be described as consisting only of MATTER/ENERGY and its inherent properties. In practice any description is incomplete due to lack of completeness of information. If this uncertainty is acknowledged it may be called secular AGNOSTICISM. In conjunction with theological AGNOSTICISM, it forms the concept of HYPOGNOSTICISM. Compare with EMPIRICISM.

SELF ← ▲ IDENTITY.

SENTIENT ← A property of a living organism that has a nervous system and can react to stimuli in reactions that go beyond the complex REFLEXes. It includes the property of self-awareness, including one's own emotions and mental modeling. An organism that perceives QUALIA, has some form of MEMORY and evidences signs of CONSCIOUSNESS. Current methods are not adequate to confidently assign sentience to borderline cases.

SET ← In Mathematics all members of a defined group. Often used loosely along with 'category' which is about a member of a set in all its varieties across other sets that have different, but related, properties.

S/HE ← Gender neutral pronoun used in this book. Pronounced 'sheh-hee". Possessive form is his/her. → MENSCH.

SHOULD ← A fraught term that implies that one desires a specific effect that requires a specific cause that "should" be used. It is part of an if – then logic that assumes the VALUE of that effect for the (implied) goal.

SKEPTICISM ← The stance that a claim of knowledge, a mental MODEL of the real world, is incomplete. I can be seen as part of the spectrum concerning world-view attitudes of Transcendent beings: Abrahamic God (Theism) – DEISM – Skepticism – ▲HYPOGNOSTICISM – AGNOSTICISM – ATHEISM.

PYRRHONISM is an early form of philosophic skepticism.

SPACE/TIME ← Qualities of reality, that in conjunction with MATTER/ENERGY, are used to describe the whole universe. They are relational properties vs. the physical properties of the latter. This relationship is summarized by Einstein's equation $E = MC^2$.

SPECIAL ← Referring to outside the ▲TYPICAL range for a situation or use.

SPIRITUAL ←A type of emotion felt when one has the experience of relating to something larger and different from oneself. To name a few, it can be felt when seeing the night sky, contemplating the size of the universe, observing large spaces of nature, wondering at the complexity of a living cell, or listening to music that one finds stirring. It fosters a feeling of peace rather than urgency. It may lead to a submergence of the self in the larger (more complex or less fathomable) entity or event. In a world view that includes the transcendent it is the feeling of the relationship between the self and any transcendent occurrence. This is a form of MYSTICISM.

STANDARD ← Accepted form for a situation or use. Synonymous with NORM. Important in the philosophy of ethics. It is not the same as TYPICAL.

STIMULI / STIMULUS ←Differentiated for clarity between <u>internal</u> versus <u>external</u> (to a body's) stimuli sources that activate sensory systems. All stimuli are forms of energy, some in conjunction with matter. As such they cause changes at the microscopic, molecular or atomic level of a cell. A nervous system speeds up the transmission of the now electrochemical secondary internal stimuli. Higher levels of nervous systems, by means of ABSTRACTION, organize multiple inputs into the next level of complexity.

STRUCTURE ←A RELATIONSHIP in space.

SUBCONSCIOUS ←All the workings of the brain that are not currently part of CONSCIOUS awareness. Subclasses are: PRECONSCIOUS, UNCONSCIOUS. → § 7.5

SUBJECTIVE ← ▲OBJECTIVE, QUALIA.

SUMMUM BONUM ← Term introduced by Cicero, meaning 'the highest good'. In all major philosophical systems, it applies the good of the individual person. Because of this it leads to frequent competing goods, which makes for relative ethics. Only a universal Summum Bonum avoids this, although in practice this scope is difficult to imagine in all its complexity. It requires each egotistical goal to be in concord with the ultimate goal that serves the most inclusive conception of value.

SUPERNATURAL ← Near synonym of TRANSCENDENT. The concept is supported by the BELIEF that a reality, outside of and different from our universe, exists. → § 5.13

SYMBIOTIC ← The mutually dependent relationship between two living things. ALTRUISTIC and PARASITIC are other forms of interrelationships. → § 14.2

SYMBOL ←An arbitrary simpler pointer to a more complex concept. It is data to a nervous system and can recall that concept if the relationship has been made in memory. The visual letters and numbers (a, b, c, ...1, 2, 100, etc.) stand for sounds. Also, verbal symbols, which in turn are concepts. In a congenitally deaf person these written symbols represent the same concepts taught by other sets of visual, tactile and other sense data. → § 4.6

T

TARDIGRADES ←" Eight-legged segmented micro-animals that are found everywhere in Earth's biosphere .Tardigrades are among the most resilient animals known with individual species able to survive extreme conditions—such as exposure to extreme temperatures, extreme pressures (both high and low), air deprivation, radiation, dehydration, and starvation—that would quickly kill most other known forms of life." (Wikipedia).

TELEOLOGY (n); **TELEOLOGICAL** (adj). ← philosophical study of PURPOSE. Much of human activity is the result of pursuing a purpose. This presupposes the ability to imagine a future. Idealism allows for non-material beings to have the same property.

THEOLOGY← The description and apologetics of gods and religion in general. It can be realistic when studying the psychologic and sociologic causes and effects of these beliefs. The instantiation of transcendent beings is, however, outside confirmational possibility.

THEORY ← a subtype of conceptual TRUTH as held by a well-informed person or group. → HYPOTHESIS and → EXPERIMENT.

Glossary

THOUGHT EXPERIMENT ←A device of the IMAGINATION. It may contain contra-factual or unlikely axioms as well as scenarios of low probability. Often use to clarify concepts.
    ← Trolley Problem; Also, is a vat of individual cells a person? → § *10.6*

TIME ← The fourth dimension concerning the measurement of rate of change that is not   MATERIAL. At the very small scales of Quantum mechanics its definition may become fraught. (Rovelli, 2018).

TRANSCENDENCE ← the proposed condition of reality that is not part of the totality of our MATTER/ENERGY universe, and which by definition cannot interact with our universe without it becoming part of it. "*In Kantian philosophy:* being beyond the limits of all possible experience and knowledge", (https://www.merriam-webster.com/dictionary) (2021). A definition that would allow such interaction *(immanence)* would open the door to a chaotic worldview.   → CHAOS.

TRANSCENDENT ← In this book replaces the use of SUPERNATURAL to indicate the concept of something outside of nature – but not "over" or "better". Having no EXISTENCE in our universe. Belief in such a reality, and their impossible impact, leads to a world view that is not confirmable by science or any other means. Any conclusions are therefore not accessible to any judgement of possibility, probability or ▲TRUTH (all capitalized). Not clearly separated from *transcendental* which is the opposite of *immanent (see above)* in the sense of non-material, *in* some theologies.

TRIAGE ← Initially a battlefield term describing the basis of the decisions concerning which causalities are given preference in treatment with the limited resources and abilities available. Its application in making choices makes the development of ethics fraught.

TRUTH ←When all capitalized refers to reality, the thing in itself, "das Ding an Sich" or NOUMENON, as described by I. Kant. In speaking it may be called "baseline truth". Truth (lower case) is based on reality in various ways. It is a quality assigned to a concept when we feel our information is adequately complete that the MODELs will give good predictions. When invoked it is an expression of BELIEF, FAITH AND CERTAINTY.
For a discussion of the derived subtypes which include the following, see Chapter VI:
    -absolute truth
    - communal truth
    - conceptual truth (knowledge) …. (Continued next page)

- logical truth
- propositional truth
- received truth

TYPICAL ← In this book used instead of 'average' or '**normal**' because the latter are often used in a pejorative way. Used to describe a range in concepts, properties, and actions. The approximate mid 80% is suggested → *4.4*. Near-synonym of the adjective "characteristic" but indicating an ordinal or cardinal scale. It indicates that we are not speaking of unusual, rare, or outlier cases. ATYPICAL is outside that range but usually carries no positive or negative value judgement. Other terms that are used to indicate a place in a range include Median; Outlier; STANDARD; SPECIAL.

* U *

ULTIMATE GOAL ← The only one with intrinsic value to which all aims must be in concordance if one is to have an ethical system where contradictions can be usefully investigated and hopefully solved.

UNCONSCIOUS ← Those workings of the brain which are never directly available to the CONSCIOUS mind. It is however part of the workings of the BRAIN-MIND that builds MODELs of reality, as well as fantasies, in a process called IMAGINATION → § *5.9* It is a subset of the ▲SUBCONSCIOUS.

UNIVERSE ← All of SPACE/TIME and its MATTER/ENERGY contents. In certain scientific theories it is part of separate, noninteracting Multiverses. Synonym is Cosmos.

UTILITY ← The property of being a MEANS (noun) to a GOAL. In Utilitarianism (U) this means(verb) that an end (goal) justifies that Means. There are many subtypes. Early proponents are <u>Jeremy Bentham</u>, <u>John Stuart Mill</u>, <u>Henry Sidgwick</u>, <u>R. M. Hare</u>, <u>Peter Singer</u>. → § 13.8. <u>Social U. is approximately summarized by the phrase "the greatest good for the greatest number". Egotistical U. is at its most extreme a form of HEDONISM. In the latter case the value is based only on human emotion.</u>

UTOPIA ← A shared concept of a perfect, ideal world. Its details are never as precise as those of an EGOTOPIA because it is a group view. → § 7.1

UUA (Unitarian Universalist Association) ← Seven Principles. As per: (<u>https://www.uua.org/beliefs/what-we-believe/principles), 2021.</u>

"As Rev. Barbara Wells ten Hove explains, "The Principles are not dogma or doctrine, but rather a guide for those of us who choose to join and participate in Unitarian Universalist religious communities."

Glossary

1st Principle: The inherent worth and dignity of every person;

2nd Principle: Justice, equity and compassion in human relations;

3rd Principle: Acceptance of one another and encouragement to spiritual growth in our congregations;

4th Principle: A free and responsible search for truth and meaning;

5th Principle: The right of conscience and the use of the democratic process within our congregations and in society at large;

6th Principle: The goal of world community with peace, liberty, and justice for all;

7th Principle: Respect for the interdependent web of all existence of which we are a part."

V – Z

VALUE ← Extrinsic value: Formula: V~ N/S: (~ means "is proportional to"). Therefore, the Value of a biologic or psychologic SATISFIER is proportional to that NEED and inversely proportional to the availability and effectiveness of the SATISFIER of that NEED: → Chapter X. For the purposes of ethics, what is only valuable for individuals, in the here and now, is too restrictive. → §§ 10.3, 10.5

← Intrinsic value: A controversial concept. It is a property of an ultimate goal, one that is not a means to further goals.

VECTOR ←A quantity having direction as well as magnitude.

VENN DIAGRAM ← Widely-used *diagram* that shows the logical relation between sets using overlapping closed areas, usually circles or ovoid. Items that share two or more characteristics will be in the same enclosed area. Concerning MEANING → § 9.9.

WIGNORANCE ←A portmanteau neologism shortening willful ignorance. The result of a combination of COGNITIVE DISSONANCE and intellectual laziness. It is a factor in the lack of integration of science and philosophy.

WILL ← The noun in FREE WILL; the purposeful tendency to action. The intensity of the will is governed by the strength of the accompanying emotions. The lack of will, is called apathy. Total apathy leads to inaction or purposeless action. → §§ *13.11-2*

WISDOM ←a product of:

- Abstraction — the structured patterns of neuronal activity initiated by responding to sensory information derived from the real world, including other neuronal circuits.

- Imagination — the forming of new MODELs based on the reconfiguration of the basic abstractions and memory.

-Memory — the repository of patterns and concepts derived from both abstraction and imagination.

- Consciousness — the real-time mental activity necessary to create and retrieve detailed memories <u>and</u> also to create the concept of a future. These necessary functions do not allow us to speak of conscious evaluations in RLTs.

→ What is wisdom? § *7.6*
→ Its purpose in Communication → §§ *8.1, 8.5*
→ As a product of evaluation → § *9.2*
→ In Ethics → § *14.9*

WORLD VIEW ← One's philosophy composed of four parts:

- Ontology, ▲REALISM;
- Epistemology, ▲HYPOGNOSTICISM;
- Evaluations, ▲ BRAIN-MIND;
- Ethics, ▲HUMANE PRAGMATISM.

Each is treated in sequential Parts (I – IV) in this book.

ZYGOTE ←A diploid cell, having two sets of chromosomes derived from the joining of two haploid (one set of chromosomes) cells in the process of sexual reproduction. Some organisms, especially plants, propagate by producing ramets (genetically identical individuals) formed by vegetative reproduction, natural or by cutting

Bibliography & Resources

Acton H.B. *"Hegel"* in "The Encyclopedia of Philosophy, editor Edwards, Paul", Macmillan 1967

Arendt, Hanna. *Eichmann in Jerusalem: A Report on the Banality of Evil*. Penguin, New York, 1994.

Audi, Robert (editor). "Preface, 1st Edition pg. xxix, *The Cambridge Dictionary of Philosophy, 2nd Edition*, Cambridge University Press, NY, 1999.

Barrow, John D. *Impossibility: The limits of science and the science of limits*. Oxford, NY, 1998.

Bassett, Danielle S. and Gazzaniga, Michael S. Understanding complexity in the human brain. *Trends Cogn. Sci.* 2011 May; 15(5) 200-209(NIH Public Access).

Beiser, Frederick. *Schiller as Philosopher, A Re-Examination*. Clarendon Press, 2005.

Boyd, Graham W. *The brains risk/reward system makes our choices, not us*. Philosophy Now Issue 112: February/March 2016.

Brown, J.F. *Psychology and the Social Order*. McGraw-Hill, NY & London, 1936.

Bunge, Mario. *Philosophical Dictionary Enlarged Edition*. Prometheus Books, Amherst, NY, 2003.

Bunge, Mario. *Philosophy in Crisis: The need for Reconstruction*. Prometheus Books, Amherst, NY, 2001.

Bunge, Mario. *Chasing Reality: Strife over Realism*. Toronto, 2006.

Carroll, Sean. *The Big Picture*. Dutton NY 2016.

Cathcart, Thomas. *The Trolley Problem or... A Philosophical Conundrum*. Workman, NY 2013.

Caygill, Howard. *A Kant Dictionary*. Blackwell. Cambridge, MA 1995.

Churchland, Patricia. *Touching a Nerve: Our Brain, Our Selves*. Norton, 2014.

Casti, John L. "*Paradigms lost. Images of Man in the Mirror of Science*". William Morrow, NY 1989.

Chopra, D. and Tansi, R. *"Rethinking our understanding of Genetics"*, (https://deepakchopra.medium.com/rethinking-our-understanding-of-genetics-c4628dc63957), 2019.

Deleuze, Gilles et al *"What is Philosophy"*. Columbia University Press 1994.

De Waal, Frans. *"Mama's Last Hug"*, Norton, 2019.

Durrant, Will. "*The Story of Philosophy*", Pocket Books, NY 1961.

Dye, Frank. *"Human Life Before Birth "*. Harwood Academic, 2000.

Eagleman, David. *"Livewired: The Inside Story of the Ever-changing Brain"*. Pantheon, NY. 2020.

Emmeche, E., Køppe, S., Stjernfelt, F. https://www.nbi.dk/~emmeche/coPubl/97e.EKS/emerg.html .(published [with minor modifications] in: Journal for General Philosophy of Science 28: 83-119 (1997)).

Flew, Anthony, *"How to think straight"* Prometheus books, 1998.

Free, Ann C. (editor), *"Animals, Nature and Albert Schweitzer"*. Washington DC, Flying Fox Press,1988.

Gaiman, Neil, *"Death: The High Cost of Living"*. *Vertigo (DC Comics)* 1993.

Greene, Joshua. *"Moral Tribes: Emotion, Reason, and the Gap between Us and Them"*. Penguin Press, NY. 2013.

Grobstein, Clifford. *"The Strategy of Life"*. W.H. Freeman. 1965.

Hall, Stephen S. *"Wisdom from Philosophy to Neuroscience"*. Alfred A. Knopf, NY ,2010.

Harari, Y.N. *"Sapiens"* HarperCollins, NY, 2014.

Haidt, Jonathan, *"The Righteous Mind"*. Vintage Books, 2013.

Hartman, Robert S. "The Structure of Value: Foundations of Scientific Axiology", (Paperback)Wipf & Stock, 2011.

Hartmann, Nicolai. *Nicolai Hartmann* in https://plato.stanford.edu/entries/nicolai-hartmann/#OntoCate, 2021.

Kahneman, Daniel. *"Thinking, Fast and Slow"*, Farrar, Straus and Giroux, 2011.

Kunzmann, Peter et al. " *Dtv-Atlas Philosophie"* Deutscher Taschenbuch Verlag, Munich. 1998.

Lane, Nick. *"The Vital Question"*. W.W. Norton, NY, 2015.

Legg, Catherine and Christopher Hookway, "Pragmatism", *The Stanford Encyclopedia of Philosophy* (Fall 2020 Edition), Edward N. Zalta (ed.), URL = https://plato.stanford.edu/archives/fall2020/entries/pragmatism/). 2021.

Mahner. Martin (editor). *"Selected Essays of Mario Bunge— Scientific Realism"*. Prometheus, Amherst NY, 2001.

MacLean, Paul D. *"The Triune Brain in Evolution: Role in Paleocerebral Functions"*. Springer, 1990.

Mackie, John L. *"Ethics — Inventing Right and Wrong"*, Penguin Books, 1977.

McDowell, John. *"Mind and World"*, Harvard University Press, 1996.

McMichael, A.J." *Climate Change and the Health of Nations"*, Oxford Univ. Press, New York, 2017.

Mead, Hunter. *"Types and Problems of Philosophy"*, Henry Holt, NY,1946.

Melchert, Norman. "*A Great Conversation, A Historical Introduction to Philosophy*", Mountain View, California, Mayfield Publishing Company (7th edition 2014).

Menand, Louis. *"The Metaphysical Club"*. Farrar, Straus, and Giroux, NY, 2001.

Mercier, Hugo and Sperber, Dan. "*The Enigma of Reason*", Harvard University Press, 2017.

Merriam-Webster.com (https://www.merriam-webster.com/dictionary) (2021)

Minnich, Elizabeth Kamarck, "Transforming Knowledge" Philadelphia, Temple University Press,1990.

Misak, Cheryl. "*Frank Ramsey: A Sheer Excess of Powers*". Oxford, 2020.

Needham, Rodney. "*Against the Tranquility of Axioms*" Berkeley, University of California, Berkeley, 1983.

Notzig, Robert. " *Philosophical Explanations*", Harvard University Press, 1981.

Passmore, John. *Encyclopedia of Philosophy.* N. Abbagnano et al (eds), USA, Macmillan, 1967.

Pinker, Steven." *The Blank Slate: The Modern Denial of Human Nature",* Viking ,2002.

Rawls, John *"A Theory of Justice*" 1971.

Rovelli, C. *"The Order of Time"* Riverhead Books (Imprint of Penguin Radom House). 2018.

Runes, Dagobert D. "*Spinoza Dictionary* ", Philosophical Library, Inc, New York, 1951.

Russell, Bertrand. "The Art of Philosophizing & Other Essays", Littlefield Adams, 1968.

Safranski, Ruediger. "*Schopenhauer and the Wild Years of Philosophy*". Harvard University Press, 1990.

Sagan, Carl, "*The Dragons of Eden: Speculations on the Evolution of Human Intelligence".* Random House ,1977.

Sallis, John. *"On Translation.* Indiana University, 2002.

Sapolsky, Robert M. *"Behave - The biology of humans at our best and worst".* Penguin Books, NY, 2018.

Schumacher, Ernst F. *"Small is Beautiful: Economics as if People Mattered".* HarperPerennial, (reprint) 1989.

Schweitzer, Albert." *Indian thought and its Development".* Beacon Press, 1936.

Schweitzer, Albert. (translated from French by J.B.Gerald). "The Difficulty of Ethics in the Evolution of Human Thought". Inkspot Press, Bennington VT. 1985.

Schweitzer, Albert. "*The Philosophy of Civilization*". Macmillan New York 1960.

Schweitzer, Albert. "*Reverence for Life*", Harper & Row, 1969.

    Ch. 15 *Reverence for Life.* Sermon 1919 pp 112-116

Ch 16 *Ethics of compassion Sermon*, 1919.
Seldes, George 1985. "The Great Thoughts". New York, Ballantine Books, 1985.
Sheldrake, Merlin. *"Entangled Life"*. Random House NY, 2020.
Shook, John R. *"Dewey's Empirical theory of Knowledge and Reality"*. Vanderbilt, 2000.
Sowell, Thomas. *"Intellectuals and Society"*. Philadelphia: Basic Books, 2010.
(SEP) –http://plato.stanford.edu/; online "Stanford Encyclopedia of Philosophy", 2021.
Thagard, Paul." *Brain-mind: From neurons to consciousness and creativity*". Oxford, 2019.
Tiles, James E. *"Dewey"*. Routledge, 1988.
The Philosophers Magazine l, Issue 50, 2010
Trusted, Jennifer. *"Freewill and responsibility"*: Oxford, 1984
UUA (Unitarian Universalist Association), *"Building Your Own Theology"*. 2021 (https://www.uua.org/products/building-your-own-theology-volume-1)
Vintiadis, Elly. *"Emergence"*. In *"Internet encyclopedia of philosophy"* https://iep.utm.edu, 2021.
"Webster's New Collegiate Dictionary". Merriam, Cambridge, Mass. 1956.
Wilson, E.O. *Consilience: The Unity of Knowledge"*. Knopf, 1998.
Wittgenstein, Ludwig. In *"On Certainty"*, eds. Anscombe, G.E.M., Von Wright G. H. Harper & Row, NY, 1969

About the Author

Dr. Koepke has extensive exposure to science in academics, research and application. He has a BA in chemistry, an MA in behavioral toxicology, a PhD in Pharmacology and an MD with a specialization in pediatrics. Although his academic training is strongest in the sciences it was medicine that provided the impetus to study ethics –including a medical ethics university course. This led to a further expansion of self-directed learning in philosophy. He found that for the most part philosophy had not kept up with the modern scientific advances, especially in the neurosciences including psychology. Born in Germany, married to a Hungarian and having many Hispanic patients caused a parallel interest in languages and their use in human discourse. This book is the result of 20+ years of developing a philosophy that is underpinned by a scientific understanding of our universe, especially our earth.

Printed in the USA
CPSIA information can be obtained
at www.ICGtesting.com
LVHW090350061024
792987LV00006B/783